THE OWNER'S GUIDE TO THE BODY

My Mission: To boldly go to the places massage does not reach.

My Purpose: To create, experience and share wholeness.

My Dedication: To the women who have guided me to adulthood, past, present and future.
Housekeeper and third parent, Marie Kirby;
my mother, Judith Anne Hornung;
my mother-in-law, Joan Marjory Jones;
my wife, Dawn Alexandra Golten;
and my daughters, Francesca Daisy and Angelica Mae.

THE OWNER'S GUIDE to the BODY

HOW TO HAVE A PERFECTLY TUNED BODY AND MIND

Roger Golten

Thorsons
An Imprint of HarperCollins*Publishers*

Thorsons
An Imprint of HarperCollins*Publishers*
77–85 Fulham Palace Road,
Hammersmith, London W6 8JB

First published 1999

1 3 5 7 9 10 8 6 4 2

A catalogue record for this book
is available from the British Library

ISBN 0 7225 3737 9

Printed and bound in Great Britain by
The Bath Press, Bath

Contents

Acknowledgements

Special thanks are due to Fiona Harrold, for being so clear that I had to write a book, Belinda Budge, commissioning editor at Thorsons for believing in it, Charlotte Ridings, my editor, for her work, Kendal Jordan, early collaborator and sounding board, and Steve, Trev and Kate for concept work on the cover and a great deal of positive support.

I would also like to acknowledge Georgiana Gore PhD., for introducing me to my body, Werner Erhard for introducing me to myself (as opposed to my mind), Darlene Fillius for introducing me to Hellerwork, Mark Stephenson for telling me about it in my hour of need, Joseph Heller for the greatest, safest space of learning I have ever known, and Lew Epstein for the space to listen with compassion and know that I am loved.

The Owner's Guide to the Body

'The greatest mistake that humanity makes is not recognising that the only way we get anywhere is by making mistakes. We're given a left foot and a right foot so that while we make little mistakes in each stride, between the two we get to where we want to go. This is the way Universe operates.'

Bucky Fuller

'There is one outstandingly important fact regarding your body and that is that no instruction manual came with it.'

(With apologies to R. Buckminster (Bucky) Fuller)

If the Manual had been written, nobody except zealots would have read it anyway, and the rest of us would have rebelled against it. Personally, I never read the instructions thoroughly before trying a new gadget or gizmo out – just enough to start me off, then it is all experiential. We tend to learn by making mistakes. The more mistakes you make, you more you learn.

Getting from here to there is a navigational problem. There are no straight lines in the real world, so progress is a process of error correction, moving towards the ideal in a sort of a zig-zag, correcting your course as you go, trying not to overdo it, always dealing with unforeseen circumstances as best you can.

The Chinese word for 'crisis', *wei-chi*, contains the characters for 'beware, danger' (*wei*) as well as 'opportunity for change'(*chi*). Bucky Fuller talked about 'emergence through emergency'. It is the way that what is vitally needed is brought powerfully to our attention so we actually do something about it: danger becoming an opportunity for change. It is up to you to find out how your body works, how to maintain it and how to get the best out of it. If you are currently experiencing musculo-skeletal problems they may turn out to be the best thing that ever happened to you. After all, would you have bought this book without the incentive of a possible solution for an issue, 'problem', or question that you may have?

This book will supply the vital information you need to understand your body better; what it is trying to tell you and what you can do about it. Aches and pains can be positive messages, if you can decode them correctly.

Remember – breakdowns can be irritating, inconvenient and expensive!

Radar was developed and used by the British during the Second World War when the need for air defence became top priority as the Luftwaffe began attacking the UK mainland in 1941. Limited resources were marshalled and expertise brought to bear to quickly produce a solution to a dangerous problem and to apply the knowledge, which already existed, in a practical way

Foreword

Dear Reader

You are embarking on a journey of discovery, or rather of self-discovery. I am excited to see this book published for its subject is a very timely one. Our ignorance about our bodies and our lack of awareness of our physical self have created conditions of extreme stress both in our culture and in our environment.

In this manual Roger has displayed the rare combination of enthusiastic attention to detail that comes from being passionate about one's subject, and of sophisticated overview that results from a keen mind. He has the ability to deal with everyday subjects through very unusual perspectives which create a more meaningful context for our body as the physical metaphor of life.

Readers of every persuasion will become more enlightened about how their world view shapes their bodies and how in turn their bodies shape their activities. Fortunately, since this process works in both directions, we can, through the appropriate viewpoint, reshape our physical structure and use our new consciousness to change ourselves as well as our world. I commend Roger Golten for making this possibility easily understandable and accessible to everyone.

Joseph Heller

Introduction

This is a book about the body written from the point of view of the 'owner/driver' – all the information is practical, non-technical, jargon-free, easily assimilable and organized for instant access. There is a separate chapter on each of the most common activities of modern Western humans – activities which are so ubiquitous as to have been completely unexamined and overlooked: the way we sit, stand and walk being the most obvious, as well as how we breathe!

The explosion of choice in lifestyles, social and geographic mobility, practical design, morality and value systems, and the advent of a new century, conspire to make it even more important and timely that we start to work from a more rational perspective in the use of ourselves.

We have applied our intellect to the world around us to massive and powerful effect but, to a surprisingly large extent, not to ourselves and or our own use of ourselves. In this book I hope to begin to redress this and initiate a process for you whereby you can find some possibilities for improving your use of yourself, and can make progress in the next (and possibly mankind's final) step in your physical evolution – just before the evolution of the body is eclipsed completely by technological development.

The purpose of this book is to make sure that you, the driver of the vehicle known as a human body, which has been rolling off the production lines in its current form for about 200,000 years, have the necessary information you need to maintain proper functioning for as long as possible. Of course, from time to time it may be proper and useful to get specialized help, but the driver needs to be in a position to determine what, when, where and how – in consultation with the 'expert' engaged. The idea then is to beef up your knowledge of, and confidence in, your body. After all, you are a walking, living example of the most complex structure discovered thus far in the entire Universe, and you are capable of understanding yourself better than any expert. Your thoughts, feelings, emotions and sense information are more complete than any MRI scan or ultrasound investigation can provide.

To take just one example. Conventional approaches to treating back pain have proved highly ineffective. There is very limited understanding of chronic back pain by the medical establishment, and very little awareness of the body's overall structure and function as a hydraulic, tensionally organized life-form. If you go to your doctor with back pain, or with many other musculo-skeletal problems, he or she is unlikely to suggest a programme of postural improvement. For the doctor, the body that walks in through the door is seen as a relatively static structure – certainly not capable of changing shape or habitual use. This structural metaphor, however, offers a rational basis for understanding and treating backache in an educational, self-help process which is quite different from that offered by gymnasia, physiotherapists, orthopaedic surgeons and others.

This is not primarily a book of technique, or rather, the techniques which are offered are done so from the point of view of a larger context of understanding the vital principles of human movement – that there is a greater field of energy in which we operate that needs to be recognized and accommodated.

This is a vital part of the Hellerwork method. Joseph Heller has been more concerned with teaching his students to see what needs to be done than to teach specific techniques. If you cannot see what needs to be done, any amount of technique will not help – this is like the plumber who charges £99 for knowing where to hit the system with his hammer and £1 for the labour. A young man once asked Buckminster Fuller what he should do with his life. Bucky replied that he should do what he sees needs doing that no one else is attending to.

Our free will, together with a vast array of technology, gives us many choices in life, and many more opportunities than before for choosing optimal responses than in earlier times. By the same token, there is more chance of making less effective choices. This book is about providing the opportunity to get more ease, balance and integrity into your primary relationships: your relationship with yourself, and your relationship with the planet you are currently residing on: Planet Earth.

> The human body is the magazine of invention, the patent office, where are the models from which every hint was taken. All the tools and engines on earth are only extensions of its limbs and senses.
>
> Ralph Waldo Emerson

ABOUT THE AUTHOR

The information contained in this book represents a distillation of my life experience to date, including 15 years' continuous full-time practice of Hellerwork. Hellerwork is a dynamic programme to improve posture, energy, vitality, presentation and confidence through a synthesis of deep tissue bodywork, movement awareness and a conversation to integrate the mind, body and spirit.

I did a BA (Hons) degree in International Relations at Keele University in the UK, graduating in 1976. At Keele I became involved in an extra-mural improvisational dance movement group, led by Georgiana Gore, a PhD student in the Sociology department, who had spent time studying in America with Don Oscar Becque, a virtually unknown collaborator of Mabel Elsworth Todd, the author of the classic work of movement study, *The*

Thinking Body (1937) and a contemporary of the modern dance pioneer, Merce Cunningham.

Via Becque, Georgiana taught an unusual combination of exercise and movement designed to improve body structure and raise consciousness as much as to develop performance skills, which we also did.

Structural exercises known as 'dwelling' and 'constructive rest', together with ideokinesis, formed the mainstay of the programme that Georgiana taught. This information remained unpublished until this book, where I reproduce some the constructive rest/lines of action information, with the permission of Dr. Gore, Becque having died in obsurity in New York some years ago.

This practice of structural exercise had a major impact on my well-being. Having shut down emotionally to cope with the trauma of my father's suicide five years earlier, I was able to get back in touch with myself through becoming aware of my body, my posture and my limitations in movement. I discovered the ability to change patterns and habits through attention, awareness and choice. By becoming more aware of the design of my body I was able to consciously choose to start using myself better; and by using myself better my body started to come out of its (unconsciously) depressed pattern.

Browsing in the university bookshop one day, I picked up a book by R. Buckminster Fuller, edited by James Meller: *The Buckminster Fuller Reader.* Fuller was discussing economics; the traditional definition of it as the allocation of scarce resources, and then went on to assert that there is no scarcity! This appealed to me greatly, disillusioned as I was with economics as it was taught, and I began a

12-year odyssey to track down and read all Fuller's books. It is a regret that I never got to meet him before he died in 1983. I have, however, visited the Buckminster Fuller Institute and met many of Bucky's family, friends, associates and students from all over the world, and since I completed my first reading of his work I have continued my research into Fuller and his work.

Fuller reintroduced me in a totally refreshing way to science, maths, physics, chemistry and, for the first time I heard about design, systems thinking, metaphysics, architecture, and many other subjects in a way that made them accessible, understandable and inspiring. His life was an example of fortitude, persistence, integrity, bloody-mindedness and independent thinking that would be hard to match anywhere, anytime. In this book I have quoted extensively from Bucky Fuller. In my opinion he is not as well known as he might be, or will become in the future.

I came to Hellerwork as a client: I was a guy with a backache. I experienced a lot of lower back pain in the early 1980s and had sessions with a visiting American Hellerwork practitioner called Darlene Fillius. It was quite a personal challenge for me to see her – I had already ducked out of seeing an osteopath, fearing being 'done to' and of being out of control, and I was sure that no-one could help me. I had a feeling I had to do it by myself, but a desperate sense that nothing could be done.

Hellerwork was a revelation – I could be helped without being disempowered! In fact it was the most inspiring hour I had ever spent. I returned to continue the standard course, and after three sessions I knew that this was something I wanted to share with others. I was enthusiastic about my new-found vocation. Within six months I was on my way to train with Joseph Heller.

Joseph Heller had begun Hellerwork in 1978. Darlene was on the second training, I found myself on the fifth. Now in 1998 there are Hellerwork trainings established on four continents and over 500 practitioners have been trained.

For myself, this experience has been about mending the split between mind and body by applying what I already knew in my head back to myself as represented by my body. I have discovered that I do not reside exclusively in either my mind or my body, but rather I am 'meta' to both, and I distinguish this way of being – where there is a quality of internal reciprocity and communication between the the mind and the body which informs, stimulates and lends creativity – from the other way of being, which is a one-way modus operandi called 'mind operating body'. This latter is not consistent with reality, it is ungrounded and leads to errors of judgement.

PROVENANCE OF HELLERWORK

More than 60 years ago Dr. Ida P. Rolf discovered that she could alter the shape and balance of a human body by a combination of deep pressure and client movement along anatomically and geometrically accurate lines. This represented a synthesis of her knowledge of yoga, Alexander, osteopathy and chiropractic, biochemistry and physiology (the latter two being her PhD subjects). She developed a uniquely potent method of actually altering the deep structure of the human form, called Structural Integration.

Joseph Heller trained with Dr. Rolf in 1972. His background was scientific, and he worked on the first communications satellites before his interest in human potential took him to the directorship of The Kairos Institute in Los Angeles. There he organized workshops and seminars for Buckminster Fuller, John Lilly (explorer in human consciousness and interspecies (human/dolphin) communications), and Virginia Satir (the renowned developer of family systems therapy), amongst others, before meeting Dr. Rolf and submitting himself to be used as a 'guinea pig' in a public demonstration. From that point on he knew he was going to study with her. In 1976 Heller was appointed first President of the Rolf Institute. In 1978 he left the Rolf Institute to begin his own teaching.

Hellerwork teaching is about creating a space which empowers clients to become identified with the values and goals of the teaching, by having them recognize that who they are is integral with what they are being taught. The experience of Hellerwork is having your body catch up with your mind, so that the baggage of the past is released and you live more in the present. Hellerwork brings harmony and balance into the body and the inner and outer relationships you have with yourself and the Universe, allowing you to express yourself more fully in the world.

It is a synthesis of classic structural integration and personal growth work, which together transforms your relationship with your body and your experience of being alive. Hellerwork will improve overall efficiency and ease in body mechanics, and help to integrate body and mind, and as such is a valuable tool for enhancing performances in sport and the expressive arts.

'Love is metaphysical gravity'
Attributed to Bucky Fuller and Oscar Ichazo

Hellerwork is a systematic, structural approach to holistic health, based on an understanding of the relationship of man with the single most powerful force operating on his body throughout life – gravity.

Hellerwork continues to evolve. Recently, the Hellerwork Board of Education was able to distill four Hellerwork Foundational Principles:

- We recognize the existence of a greater field in which we live, interact and express ourselves.
- Our purpose is to enhance the individual's awareness of and relationship to that field.
- Within the context of a healing relationship, we work with structure, psyche and movement to improve function and well-being.
- Our process follows an ordered sequence organizing the body along the line of gravity, through guided touch and education, introducing change towards a more functional pattern.

POSTURE

Posture comes from the Latin word 'ponere', which means to place. Posture implies that the superficial structure is being held in a position which is disharmonious with the core structure. In other words you are standing up straight when you would rather be slouching! How we hold ourselves, the position that we take, the posture that we adopt, all change the way we experience life and the way that life experiences us. Depress your chest, elevate your shoulders, poke your neck out forwards and then lift up your chin. How does the world look from here? How would someone else experience you in this state?

I would like to underline a big difference between 'normal' and 'average' posture. 'Normal' is something rarely seen. Normal would include a feeling of lightness and ease arising from being organized around a more vertical axis, with relaxed shoulders and a head that balances on a vertically oriented neck. Normal would include the ability to sit, stand and walk for long periods without fatigue. Slouching, paunchiness, round-shoulders, flat feet, backache, neck and shoulder tension are symptoms of 'averageness', and a structure in collapse. Normal has been summed up as 'excellence with ease'. How do you get from average to normal? The symptomatic approach just won't work here: we have to start looking at ourselves as whole systems, and treating ourselves as whole systems.

REASONABLENESS VS UNREASONABLENESS

How long are we going to put up with a world that demands conformity to the demands of mass production?

You cannot sit in that old office chair, with the seat tilted back and the arms too low, at the desk that is too high, with the telephone on the wrong side (because the cord isn't long enough), not to mention with the computer

'The reasonable man adapts himself to the world: the unreasonable one persists in trying to adapt the world to himself. Therefore all progress depends on the unreasonable man'

George Bernard Shaw

keyboard at an inappropriate angle and the VDU set too low and at the wrong angle, and expect to enjoy a relaxed and stress-free lifestyle! What's the point of loading your body with unecessary stress? There's enough of it about already. Can you really do your best work mentally whilst doing your worst physically?

This reminds me of the Parable of the Saw. A man goes into the forest with his new saw and fells 10 trees on the first day. After three weeks his friend visits him. He is only managing two or three trees a day by this time. The friend asks: 'Why don't you sharpen your saw?' The man replies, 'I haven't got time to sharpen my saw, look at all these trees I have to fell!'

In contemporary society, looking good has become more important than feeling good. This explains phenomena such as uncomfortable chairs, cars, shoes, cosmetic surgery, pollution and *Hello!* magazine. Superficial values obscuring deep needs. Struggle and effort have become desirable values in our topsy-turvy culture. People expect to suffer, life is hard and 'no pain, no gain' has become the mantra. Feel the burn! Look at the suffering in the gym and in the aerobics classes!

> The idea that one could learn to 'do' one's body differently does not occur to most people. Hence there has been a very small market for transformational movement education. Perhaps the increasingly sedentary nature of western culture must reach some critical mass before this trend changes.
>
> Mary Bond

What if you lived your day-to-day life practising excellence with ease. Would you need to spend hours each week atoning for your sins? What if you could interpret and accommodate the signals that your well-designed body sends you? What if that back/neck/shoulder/leg/foot/stomach/headache that plagues you was an intelligent message from another universe ... your body.

Engage yourself fully in this book; body and mind. Try out the little thought experiments, the awareness exercises and the new ways of doing things. I believe that if we sat, stood and walked more easily and effectively, all sorts of problems would clear up, just in the process of day-to-day life. Don't take my word for it. See if it works *for you*.

'The Doctor of the future will give no medicine, but will interest his patients in the care of the human frame, in diet, and in the prevention of disease'

Thomas Edison

'If Humanity does not opt for Integrity, we are through completely. It is absolutely touch and go. Each one of us could make the difference'

Bucky Fuller

SUMMARY

- It's OK to make mistakes – that's how we learn.
- You are a vital part of the whole.
- If you are 'trying' to have good posture you've missed the point.
- 'Get out of your head and come to your senses' (Fritz Perls).

Chapter 1

Down to Earth

One small step for man, a giant leap for mankind

Neil Armstrong

A study of the human body as a whole system cannot begin without considering the environment within which it exists. We will consider our daily surroundings, but one of the most awesome gifts mankind has created for himself is a photograph of the Earth from space, encapsulating the image of the 'Whole Earth', a symbol of our connectedness here on this beautiful, watery planet.

It was Buckminster Fuller who coined the term 'Spaceship Earth' in 1951 and was talking about the environment in the 1930s. Fuller was a true polymath, a man not constrained by traditional academic boundaries and as such, can be considered one of the true pioneers of holistic thinking. By returning to first principles he felt able to explain even the complexities of nuclear physics to children.

Fuller had a definition of Universe that included man's thinking; the metaphysical as well as the physical:

Environment to each must be
all there is that isn't me.
Universe in turn must be
all that isn't me, plus me.

Each of us is an indispensible part of Universe – we are not passive observers, we all have a part to play.

Fred Hoyle, the brilliant cosmologist, anticipated the impact of the first Apollo space mission photographs as far back as 1948. (It was also Hoyle who named the hypothesis of one explosive instant of creation as the 'Big Bang', and the name stuck! He prefers his own theory of 'Continuous Creation'.)

> Once a photograph of the earth, taken from the outside is available ... a new idea as powerful as any in history will be let loose.
>
> Fred Hoyle

'Human consciousness will be transformed by us seeing ourselves from space'

Arthur C. Clarke amplified the same prediction: 'Human consciousness will be transformed by us seeing ourselves from space.'

For the first time, the average human being could see beyond the normal earthbound perspective and view his larger environment with a new expansiveness. From the moon, the Earth looks four times as big, and five times as bright, as the moon looks from the Earth. The brightness of the Earth is due to the reflection of a proportion of the sunlight falling on its surface, the remainder being absorbed by the biosphere. (The amount of reflection – known as the albedo – is currently 0.37 (visual geometic albedo). Back in 1976 it was 0.39. (A perfect mirror would measure 1.00, i.e., 100 per cent.)

SPACESHIP EARTH

Let's take a journey through space on our own spaceship – Spaceship Earth. An Earth day is determined by the time of a single rotation of the Earth around its own axis – about 24 hours. Now remember that the circumference of the Earth at the Equator is about 25,000 miles (40,200km) so, although we live as if we are on an unmoving and relatively flat surface, the truth is (and it's easy to calculate: 25,000 ÷ 24), that the Earth is spinning at about 1,000mph (1,070kph) at the equator, and about 600mph (965kph) at the latitudes where most people on the planet live.

An Earth day is determined by the time of a single rotation of the Earth around its axis.

To personally experience the Earth's rotation, and to get aligned with this reality, Bucky Fuller suggested standing with the setting* sun over your right shoulder, so that you face east, the direction of spin of the Earth, and as the sun drops below the horizon behind, you may be able to attune yourself to the incontrovertible fact of this rotation.

It is the spin of the Earth that causes it to bulge at the equator, so that its diameter here is 27 miles (43km) greater than that at the poles. This is known as *oblateness* – the degree to which an object is less than a perfect sphere. The Earth's oblateness is 1/289th. Our spinning, bulging Earth is also travelling around the Sun at an average orbit distance of 92 million miles (148 million kilometres), a nice safe distance from the nuclear reaction that provides our heat and light. An Earth year is determined by one complete circumnavigation of this track. To find out how fast the Earth is travelling relative to the Sun, divide the distance of the orbital track by 365.25 days. The answer, to save you time and trouble, is over 67,000mph (107,800kph), or 18.51 miles per second (29.79km/sec)!

If the radius is 93m miles, the circumference, $2\pi r$, is 2 x 3.142 x 93m = 584,412,000 miles, divided by the number of hours in a year, 365 x 24 = 8,760 = 66.714; the difference from the actual is accounted for by the fact that the Earth's orbit around the sun is not circular, but elliptical

Our Sun, along with the rest of our solar system is orbiting the centre of our galaxy at approximately one million miles per hour, and the galaxy is moving away from all other galaxies at a still higher speed. Feeling dizzy yet?

YOU AND UNIVERSE

To get a realistic sense of the relative size of you and the Earth in the context of Universe, let's take off on another imaginary guided tour of the greater environment that man has so far discovered he is in. The highest mountain we have, Everest, is about five miles (eight kilometres) above sea level, and the bottom of the deepest ocean trench, the Marianas Trench, is five miles below the sea. In relation to the diameter of the Earth, this total maximum variation of 10 miles (16km) is only 1/800th. Absolutely undetectable from space: a 12-inch (30cm) diameter machined steel ball is likely to have surface variations proportionately as large.

Viewed from the Earth, the Sun has an apparent diameter of one inch (2.5cm), even though the Sun has a diameter 100 times that of the Earth. You cannot see 1/100th of one inch with the naked eye. Our Sun is a very small star: Betelgeuse for instance, a relatively near neighbour in the constellation of Orion, has a diameter larger than the diameter of the Earth's orbit around the

Sun – 186 million miles (300 million kilometres). There are 100 billion stars in our galaxy and there are 1 billion galaxies within the view of the 200-inch reflector telescope at Mount Palomar. Ninety-nine per cent of the known 100 quadrillion stars are quite beyond the range of the naked eye. So when you look up at the sky on a clear night, you are seeing about one per cent of what we know to be out there.

Despite all those billions and billions of stars and galaxies out there in space, and despite all our searching and listening, there is still no sign of other life in the Universe, although that certainly doesn't mean that it does (or did) not exist. It is water that is the vital precursor of life as we know it. Water freezes and boils within a narrow temperature range. In fact, if ice wasn't less dense than water, any ice that formed would have sunk to the bottom of the sea millions of years ago, our water would have frozen, and life would never have developed.

> Water is strange. It is the most abundant liquid in the world … Water is a scientific freak. It has the rare and distinctive property of being denser as a liquid than as a solid, and is the only chemical compound that can be found naturally in solid, liquid or gaseous states. It is also a powerful reagent, capable in time of dissolving any other substance on the planet. Nothing is safe from it, and yet this unique substance is colourless, odourless and tasteless.
>
> Lyall Watson, *The Water Planet*

Recently water, or rather ice, has in fact been discovered frozen into underground rocks around the poles of our

Water gains and loses heat more slowly than any other widely available substance on the planet. On a 12-inch (30cm) globe the depth of the world's oceans would proportionately be 3/1,000 of one inch (2.5cm) - far thinner than the absorbtion depth of the blue ink on the paper cover of the globe upon which the maps are printed, and about as deep as the condensation from your breath on to a steel ball of the same diameter

moon, and on Europa, Callisto and Ganymede – three of the Galilean moons of Jupiter. The NASA probe, Galileo, has found that Europa is entirely covered with a 60-mile (100km) thick glacier of ice, and vast oceans of liquid water may lie below that. Ganymede has a thin water icecap on one of its poles and Callisto is an icy, rocky crater-covered moon.

The European Space Agency's infrared space observatory has been finding frozen water and water vapour all over the Universe in the three years since it was put into orbit, but Earth remans the only place we know of with large liquid oceans of water.

The water on our moon will be enough to support a colony there for some hundreds of years and will provide a vital stepping stone for the exploration of deep space, allowing rocket fuel to be manufactured there. This will spare rockets from lugging it into space from the hard-to-escape gravitational pull on Earth.

Now you know where you are. What now? Perhaps you feel humbled by the immense size of Universe and our own insignificance, but remember that it is only through the success of the curious, thinking mind of man that any of this is known at all.

THE GRAVITY OF THE SITUATION

The biggest force acting on your body over the whole of your life is that of gravity – the mutual inter-attractiveness of objects or masses, the same force that holds the moon and the Earth in their orbits, an invisible tensional force that holds us onto the Earth's surface, along with the air we breathe, the water we drink and everything else.

Mysteriously, gravity is not a force that exists in its own right. It only becomes apparent as a dynamic of the relationships between things, and it quadruples every time the distance between the objects is halved. The strength of gravity is evidenced by the effort and power needed to achieve even a low-earth orbit, such as that which the Space Shuttle achieves (a mere 125–250 miles (200–400 kilometers), again undetectable when viewed from space). The escape velocity needed to leave the Earth's gravity averages out at seven miles (11km) per *second*. When you realize that this is 25,200mph (40,500kph), it is clear why some very enthusiastic, even obsessive private rocketeers have not yet been able to claim the million-dollar prize for launching something (anything!) into orbit.

'Nothing is quite so prominent in a child's life. His mother is not always around but gravity is always there. And every time he tries to stand up – Boom! Down he goes again'

Bucky Fuller

> The physicists tend to describe gravity as a 'weak' force … gravity is omnipresent, the most subtle of the great integrities … this is typical of the semantics of scientists with their addiction to axioms and solids; for the physicists a strong force is a tiny thing like a bomb.
>
> Bucky Fuller

The mass of the Earth is 6,600 million million million tons. How much do you weigh? To operate successfully on planet Earth you had better be working *with* gravity than fighting against it, and you are well designed to do that.

'Little man is so small, and his Earth is so big, that he does not realise that when he steps this way, he's pushing the Earth the other way'

Bucky Fuller

EXERCISES

Grounding 1

Sit up on a standard chair, preferably flat-seated, with your legs uncrossed and feet on the ground. Close your eyes and notice your contact with the ground through your feet and your bottom, the sitting bones. Visualize your spine extending down to the ground and into the Earth, as the third leg of a tripod of support for your body, connecting you and supporting you. Extend your tripod of support further and further in towards the centre of the Earth.

Grounding 2

While standing, open the circuit of your energetic relationship with the Earth by imagining a flow of energy up one leg and down the other, crossing from one to the other in your abdomen, like an arch, and then back into the ground like an electrical circuit.

SUMMARY

- Earth is a tiny, spinning speck in the Universe, hurtling around the Sun at vast speeds.
- The earth is the only place we know on which water – that most vital of substances – exists in its liquid state.
- Gravity is the most powerful force which acts upon us our whole lives.
- It is far simpler to conduct our lives working with gravity than against it.
- By applying the mind back to the body, we can discover more ease, balance and integrity.

* Fuller did not use the terms 'sunrise' and 'sunset', as they are not truthful. In 1934 he offered a prize to anyone who could invent new words for them: William Wainwright of Cambridge, Massachusetts came up with 'sunsight' and 'sunclipse'. The Sun does not rise or set, it is you who is moving and your personal horizon, determined by your position on the Earth, is either revealing or obscuring your view.

Chapter 2

Waterworld

Water is the *sine qua non* of life. Every cell in the body has a water content essential to its functioning. Cells cannot live without water, toxins in the body cannot be sufficiently removed without water. If you are thirsty you are already dehydrated. The developing foetus floats in amniotic fluid in the womb. Babies are moist, watery and flexible.

'Water – the ace of elements. Water dives from the clouds without parachute, wings or safety net. Water runs over the steepest precipice and blinks not a lash. Water is buried and rises again; water walks on fire and fire gets the blisters. Stylishly composed in any situation – solid, gas or liquid – speaking in penetrating dialects understood by all things – animal, vegetable or mineral – water travels intrepidly through four dimensions'

Tom Robbins, *Even Cowgirls Get The Blues*

HUMAN STRUCTURE

There is a network in the body, not generally appreciated, of fascia, or wrapping. Fascias are moist, loose, slippery, elastic in all directions, and facilitate movement under 'normal' conditions. Every muscle fibre, every bundle of fibres, muscle itself and groups of muscles, every ligament, tendon, every nerve, organ and bone, is wrapped in its own fascial bag. Even the brain and central nervous system are suspended by exquisite specializations of fascial structure: the meninges, the tentorium cerebelli and the falx cerebri.

There is only one fascia, superbly involuted, complex, capacious and capable of adapting to need, but still unified and unifying all the other body parts. Like a car body, it contains the parts, it is the context of the parts, and all the same parts in a different body create a different 'model'. If there was a magic solvent that could dissolve the body and leave only the fascias, we would be left with a perfect representation of the whole body in every detail, but without any content.

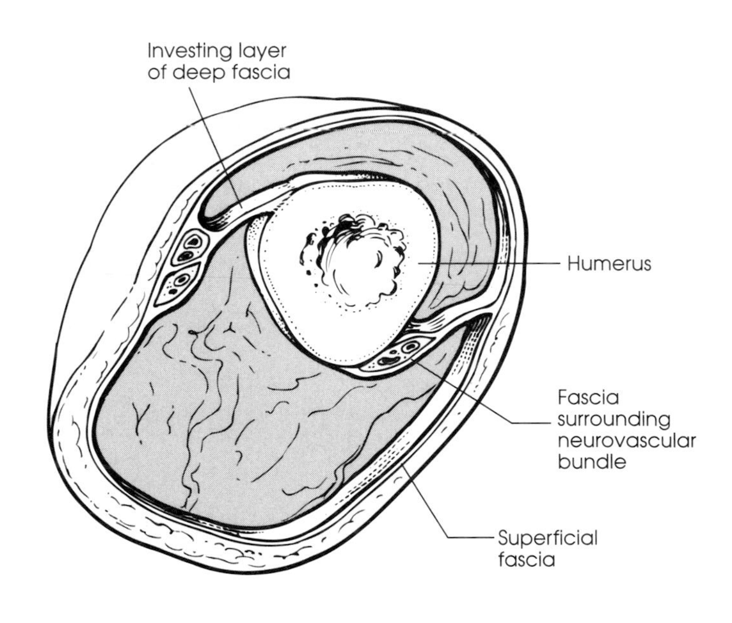

Fascias

Fascias are organized by stress, so that, for instance, if the body is sitting at an uncomfortable workstation all day, the fascias will begin to adapt to those conditions, carrying weight for bones which are out of alignment, taking the strain for muscles which become fatigued. Stress organizes the colloid, the gel-like internal structure of the fascia, into specific directionalities, and the gel begins to set, harden and dry out if the demand continues.

Adaptable, the fascias support you in whatever you do. If you sit poorly, stand poorly and walk poorly, your myofascial system will support you so far as it can. The more you train it in one direction the more it will adapt itself to meet that demand. One day you will reach up or bend down and discover that your body has lost its general adaptability. It is now only good for the limited repertoire of activities that you have trained it in. This is fine in a static world where you live in the same house, sit in the same chair, do the same job and relate to the same people, in the same way. By the same token it is something approaching fatal in the changing world of today.

THE FLUIDITY OF THE BODY

The shift to looking at the fascial network also requires a shift to looking at the body through fluid mechanics rather than solid mechanics. Water is the medium of life. Without water, no life form can survive. The body is primarily a fluid mechanism where the flow of fluids occurs at the cellular level (through osmotic and hydrostatic pressure), and at the extracellular level, mostly in the interstitial spaces. Most of the fluids of the body flow through the fascia; this means that the health of the body is related to the degree of fluidity in the fascia. With the exception of the arterial half of the cardiovascular system, the flow of fluids has no pumping mechanism other than the movement of the body itself. Water is life, movement is life. Fluidity is associated with youthfulness; rigidity with age.

'Human structuring and integral organic equipping is over 60% water. Earthians are hydraulically designed technologies. More than 50% by both volume and weight of the average physical structure and mechanics of all biological species consists of water'

Bucky Fuller

Potentially there is tremendous fluidity possible within human structure. Tensegrity is a term coined by Fuller from the words 'tension' and 'integrity' – in a tensegrity the forces of tension and compression are separated out from one another quite clearly and the integrity of the structure is maintained by a continuous network of tension, and tensegrity structures absorb stress and distribute it evenly throughout. They are very efficient in terms of weight/strength ratios.

TENSIONAL INTEGRITY

Traditional anatomical studies tend to give a view of the separateness of all the parts of the body, due to the powerful, analytical method of dissection with its contingent naming of parts (i.e., let's take this to pieces and see how it works). The view of the whole is lost in the avalanche of apparently disconnected data. Looking at this from a systems perspective, the same data yields a picture of the seamless connectivity of every fascial bag with every other adjoining fascial structure. This picture can be characterized almost as the photographic negative of the traditional understanding: the integrity of the human structure is maintained by the continuity of the fascias, the inbetweenness that was for years considered superfluous to the system.

Most bodies are compressed and their owners complain of tension. In engineering terms compression is the force of pressing together, and tension of stretching apart, and in reality these forces always and only co-exist, as mirror images to one another. Therefore you could say that tension is caused by compression, and vice versa. The human structure has tension designed into it, because tensional structures have a better performance-to-weight ratio, and most materials, including organic, biological stuff like bones, fascias and muscles are stronger tensionally than compressionally. The other compression element, as opposed to designed-in compression, comes from living in gravity but being somewhat misaligned with it.

The amazing capacity of the human body to absorb and distribute stress, and the ability to adapt to differing demands, can be better explained by using the 'tensegrity' model of structure that any other I have seen. This is quite the opposite of most structures we might be familiar with – such as most buildings. The human structure is a fantastically complex tensegrity organized around a vertical axis. Tensegrity models made with sticks and elastic show the principle; it is as if the sticks are suspended in mid-air, our eyes not accustomed to seeing tensional forces at work. None of the sticks touch one another, although the model can be very strong.

simple tensegrity model

The mental stereotype of a skeleton cannot, in reality, exsist without its envelope of soft tissues: ligaments, tendons and myo-fascia which bind everything together. Without wires and fixings the skeleton in the medical school is just a pile of bones on the floor.

In the tensegrity model, the integrity is maintained by a continuous pattern of tension in strings threaded around the sticks, which act as spacers to keep the strings apart. Without the strings, the tensegrity model is just a pile of sticks on the floor. Conceptually, this tensegrıty model is more accurate than the skeleton, because it shows the tension as well as the compressional part. The bones are only a part of the picture, and not in themselves capable of holding the structure together.

There is a Victorian quality to our common understanding of human structure. It is the world of Isambard Kingdom Brunel – magnificent, massive structures such as cantilevered iron bridges where the harnessed forces are visible and identifiable.

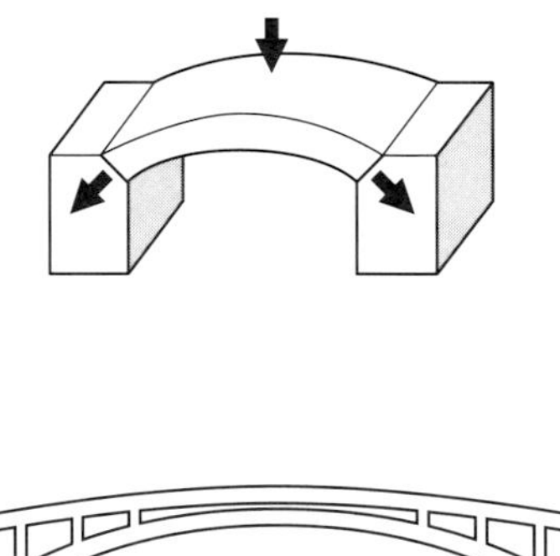

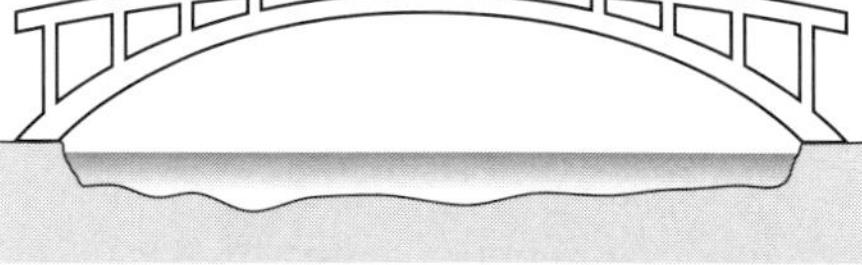

Brunel iron bridge — compressional forces

The largest clear-span structures we had until 1950 were the domes of St Peter's basilica in Rome, which took 40 years to build, 500 years ago and had a diameter of 50 metres, and the Pantheon, also in Rome, and 2,000 years old; each was made of marble and weighs approximately 30,000 tons (30,480 tonnes). Contrast the spidery lightness of Richard Rogers' Millennium Dome – enclosing 100 times the volume of the dome of St Paul's, with a fraction of the weight of materials.

Tensegrity is ubiquitous in nature yet invisible to the eye, a bit like gravity. Materials are stronger in tension than they are in compression, bones included, and that is why Nature uses it in all its myriad and fully triangulated structures.

Lord Rogers' Millennium Dome has a diameter of 350 yards (320 metres), and a circumference of 1,099 yards (1,005 metres). The total weight of this dome, which is eight times the diameter and many times the volume of the classical domes, is 1,700 tons (1,728 tonnes), less than 1/17th of the domes of antiquity. Of this total, 1,180 tons (1,200 tonnes) are accounted for by the large steel masts from which the cabling of the roof is suspended

The absence of vivid images of tensionally supported structures reduces our ability to experience ourselves as such structures. Models of tensegrity are not common in the visible, everyday world. Most structures we see are primarily compressional: bricks pressing down on one another in an arched structure, roofs pressing onto walls, etc. Some structures are partly tensional, for instance ridge-pole tents, suspension bridges and TV masts, but all depend crucially on compressional poles, pilings and bearings. Humans, however, have tensional integrity – you can stand on your head without falling apart.

The fascial network is indeed a continuous and fantastically complicated tensional system, and the bones and organs act as the compressional element. There is also an important *hydraulic* element in our structuring, taking advantage of the facts that we are 60 per cent water and that fluids cannot be compressed, or rather, that they are very good at resisting compression so that if fluid is contained it can operate as a compressional bag, maintaining the span of the tension element. In this tensegrity model the bones are the sticks and the fascias are the elastic bands which together create the structure. None of our bones touch each other (if they did we'd be in a lot of pain). The compressional and tensional elements combine to allow the range, fluidity and grace of movement that human beings are capable of. None of the robots so far built can begin to approach these qualities.

A tree is a great example of the way that nature uses fluid hydraulics to build structure. Take two two-gallon (nine-litre) buckets of water, one in each hand, and try and hold them out horizontally at arm's length. Almost impossible for more than a few seconds. Now look at the branch of a tree with the same girth as your shoulders when your arm is extended and muscles flexed. It could weigh 10 times what you can hold out horizontally – branches can weigh up to 5 tons (5.08 tonnes) or more, held out horizontally; a design accomplishment exceeding that of the 'wing roots' of jumbo jets, and also capable of withstanding great winds, flexing gracefully, even waving wildly and rarely breaking off unless dying and drying out. Great trees have been seen to bend and touch their crowns to the ground and return to vertical unscathed in hurricane conditions. As Fuller comments: 'How can a tree do that? Biological structures cope hydraulically with all compressional loadings.'

The human body is a structure that has to operate in the field of gravity, and clearly there are better and worse, more effective or less ways of carrying oneself around. Think of your head and shoulders as a superstructure that needs the support of the torso and legs – an undercarriage – to balance on. The typical head and shoulder/arm assembly weighs about 20 per cent of bodyweight, so a 140lb (63.5kg) person is carrying 28lbs (12.7kg) of superstructure around with him or her. That, as I like to say, is a lot of potatoes!

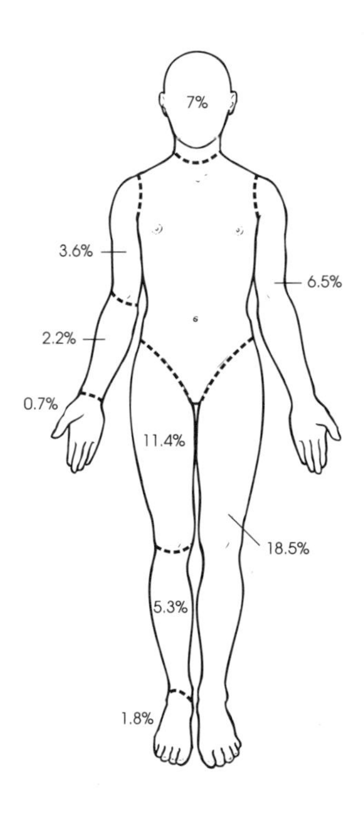

percentages of parts to total bodyweight

It is not always clear that the human body is designed to be vertically organized, and the significance of this fact can be lost as we struggle through modern life. We are losing the ability to centre on our beautiful vertical axis, with a corresponding loss of the ease and balance that is inherent in the design. Too much energy is tied up in just holding ourselves together; soft tissues are doing weight-bearing work and becoming more bone-like – hard and rigid – instead of merely stabilizing the system in its 'designed-in' equipoise. If you've ever wondered why you feel so tired after a hard but completely sedentary day at the office, maybe your body could benefit from becoming better organized structurally. However, keep in mind that the organizing line of gravity must not lead us to military rigidity in our posture.

All structures combine elements of tension and compression, they always and only co-exist. The distinction is that some structures cohere primarily on account of compression – like the brick wall – and some primarily on account of tension – like you. When you experience chronic back pain, it is often a result of compression caused by the less-than-optimal organization of the structure of your body. You are falling into compression – which is why lying down will nearly always relieve the pain.

'When we are holding our chests up with the back of our neck the whole body suffers'

Mabel Elsworth Todd, *The Thinking Body*, 1937

WATER FOR HEALTH

Dr. Fereydoon Batmanghelidj was born in Iran in 1931, went to school in Scotland and studied medicine at St. Mary's in London. He returned to Iran to improve medical care there, got caught up in the revolution of 1979 and was sentenced to death. Because of his medical background he was spared and put to work as the prison medical officer. With no medicine for his unfortunate patients, he successfully treated 3,000 fellow prisoners with nothing but adequate hydration – water.

In his book, *Your Body's Many Cries for Water*, he makes the startling proposal that many of us who live in industrial societies are chronically dehydrated, despite the ready availability of water, and that this dehydration is responsible for much physical pain and suffering. He asserts that simply drinking eight or more medium-sized glasses of water per day, together with some salt, will prevent or reverse many painful conditions such as arthritis, heartburn, back pain, asthma and many others, explaining the role of water metabolism in hypertension, headaches and weight conditions.

Dr. Batmanghelidj's theory is that our modern lifestyles tend to create chronic dehydration because we substitute coffee, tea, alcohol and carbonated beverages for the pure water that our bodies need. All of these drinks contain dehydrating agents, so the net effect on the body is a loss of water, rather than a gain. In addition, many act as diuretics, which makes matters worse.

> When the human body developed from the species that were given life in water, the same dependence on the life-giving properties of water were inherited. The role of water itself in the body of living species, mankind included, has not changed since the first creation of life from salt water and its subsequent adaptation to fresh water.
>
> Fereydoon Batmanghelidj

AQUATIC APE HYPOTHESIS

Elaine Morgan, the bestselling author of *The Descent of Woman* (1972), became more and more interested in human evolution after reading the late Sir Alister Hardy's

1960 paper 'Was Man More Aquatic in the Past?' in *New Scientist*. Anthropologists tend to discount the importance of water in the development of homo sapiens. Hardy/Morgan's theory attempts to explain the puzzling and major differences between us humans and our so-called 'cousins', the primates. The very fact of our upright, two-legged verticality is better explained, she says, by a need to wade through water rather than by ideas about migrating from the trees to the savannah, so beloved of anthropologists. Other unique features – our relative hairlessness, our downward pointing nostrils, salty tears, subcutaneous fat and constant need for water – are easily explained by this fascinating and simple idea.

> Humans, even without exercise and in temperate climates, have to drink much more than any other terrestrial mammal ... without intervention, a dehydration of about 10% may be fatal for humans, whereas most animals can rapidly recover from a dehydration of 20%...
>
> They [humans] have abundant sweat and tears, rather saturated expiration and dilute urine, watery faeces, a low drinking capacity, a naked skin, a rather thick subcutaneous fat layer, a rather low body temperature and a small circadian temperature fluctuation. Each of these features suggests than man evolved in an environment where water was permanently and abundantly available.
>
> Marc Verhaegen, 'Human Regulation of Body Temperature and Water Balance', quoted in Elaine Morgan, *The Aquatic Ape Hypothesis,* 1991

Bucky Fuller had a 'speculative prehistory' scenario, which suggests that South Pacific and North Indian ocean atolls were the 'most logically propitious places' for early humans to have survived and prospered

EXERCISES

The Fluidity of the Body

This exercise will allow you to experience the strength of your body's fluidity. You will need another person to help.

Stretch out your arm in front of you and try to keep it straight whilst the other person tries to bend it down.

Now visualize a torrent of water running through your arm, as if it were a firehose.

The other person tries again to bend your arm.

Swimming Exercises

When swimming, break up the routine by doing one or more of the following:

1) See how few strokes you can complete a length in. In breaststroke, for instance, extend the glide phase after the leg kick and before the next arm stroke.
2) Develop a dolphin kick in breaststroke. After the

normal leg kick, with your legs together try one or two dolphin movements with the lower body before the next arm stroke and breath. This does take some puff!

3) Practise swimming underwater using a dolphin movement, arms by your sides. Start with a push off the side or a dive and breaststroke arms.
4) Buy some swimming goggles and spend some time getting used to them; they will make a big difference if you haven't used them before. Stop holding your head out of the water – it is too much effort and prevents you getting the benefit from swimming. Think about your technique and your body symmetry. A little coaching/feedback goes a very long way in swimming, even if it is just a friend observing you from the side and saying what they see.
5) If you are prone to backache, breaststroke is not recommended. Develop your backstroke and frontcrawl instead. You may find it difficult at first but persevere. The breathing rhythm in frontcrawl takes practice before it can be maintained over distance – slow down the arms and speed up the leg kick. Try bilateral breathing; breathing every third stroke in frontcrawl, so you breathe both sides, or change sides on every length so that you turn your head both ways.

(Steven Shaw has a lot more about swimming and its benefits in his book *The Art of Swimming*.)

SUMMARY

- Overall, the human body is 60 per cent water; the brain 85 per cent, living bones 50 per cent.
- The body is held together by a network of moist, elastic fascias, which maintain the integrity of the structure.
- Fluidity is the opposite of rigidity.
- Life is a process of drying out.
- Drink a *minimum* of four pints of pure water per day.
- 'Tensegrity' is a mechanism that absorbs stress and distributes it evenly through the body – like shock absorbers on a car, only a lot better.
- Your body works better under tension than compression.
- The structure of your body is an elegant, complex tensegrity organized around a vertical axis.

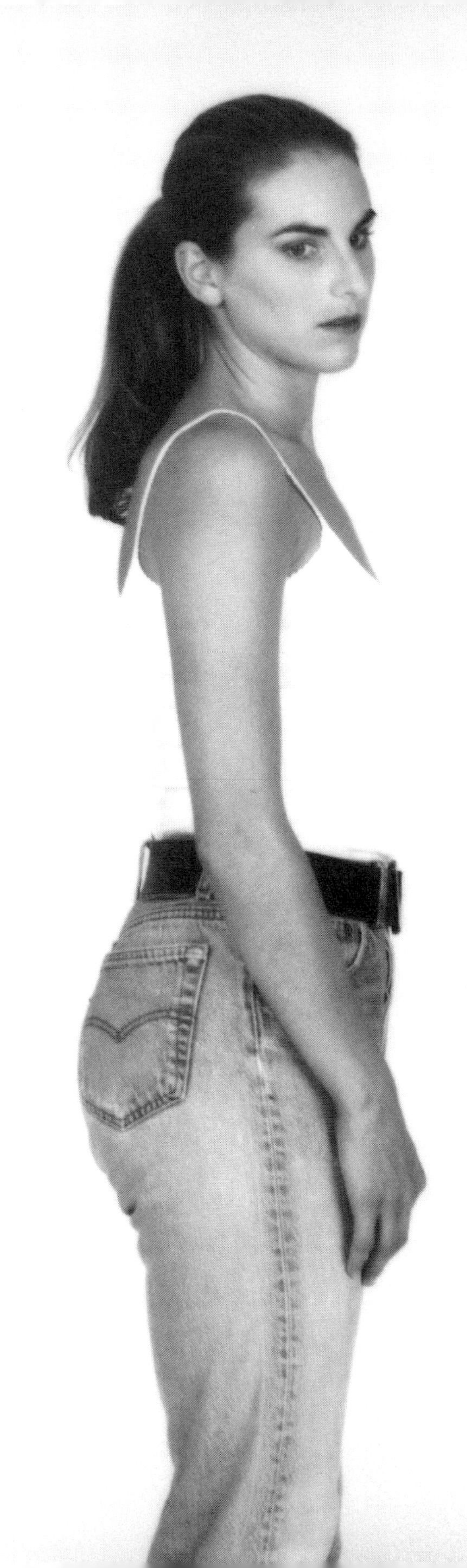

Chapter 3

'Hakimi' – Who Are You?

What a piece of work is a man! How noble in reason! How infinite in faculty! In form, in moving, how express and admirable! In action how like an angel! In apprehension how like a god!

Hamlet II, 2

'Hakimi' is a Hopi Indian word, which is currently used to ask 'who are you?' Its archaic definition is 'how do you stand in relation to these many realms?' In another spelling, 'Hakomi', it is the name given to a body-oriented psychotherapy developed by Ron Kurtz, the author of *The Body Reveals*.

The question 'who are you?' invites a range of replies covering mind, body, emotions, purpose, function, commitment, relationships and religion. It is a question that relates partly to the belief system of the individual, but it lacks a referential index: in relation to what? to whom? Many of us, if asked *who* we are, will answer the question as *what* we do: I am an engineer, a bodyworker; and our well-being fluctuates according to our success at work.

Some people will answer in terms of their relationships to others: I am so-and-so's husband or wife, etc., and their well-being may vary according to the status of their significant other.

Albert Einstein responded like this:

> A human being is a part of the whole called by us 'Universe' … a part limited in time and space. He experiences his thoughts and feelings as separated from the rest … a kind of optical delusion of his consciousness. This delusion is a kind of prison for us, restricting us to our personal desires and to affection for a few persons nearest to us. Our task must be to free ourselves from this prison, by widening our circle of compassion to embrace all living creatures and the whole of nature in its beauty.

Brugh Joy, a former cardiac surgeon turned metaphysical teacher and very much in harmony with Einstein's equation $E = MC^2$, stated that:

> I believe that the human body is an outrageously ingenious demonstration of the power of consciousness to turn energy into matter and matter into energy. With this insight we can now undefine ourselves and stop defining our limitations.
>
> Brugh Joy, *Joys Way*

ARE YOU YOUR BODY?

Some people have paid up to $120,000 to have their bodies drained of all blood and hung upside down in a freezer, so that at some point in the future they can be brought back to life, when technology allows. All, of course, after they die. Are you that kind of person? This represents an extreme form of materialism, where a person believes they are their body, and that there is nothing else, since it is unlikely that your immortal spirit would hang around to get back into the same body in some hundred or so years' time. If you're interested in this but don't have the funds, then your head alone can be preserved (for a lower fee), and as a head transplant has already been achieved with monkeys, this option is not as far-fetched as previously thought. Varieties of belief regarding the existence of an immortal spirit which animates the body range from this ultra-materialist cryogenic body-preserver position, to the spiritualist who discounts the importance of the body at all.

The human body is certainly an amazing piece of engineering. Have you ever really considered the range of functions and faculties that your body has and provides for you? In a whirlwind *tour-de-force*, Buckminster Fuller gives us a run-down on all our operating features...

> A self-balancing, twenty-eight jointed adapter base biped; an electrochemical reduction plant, integral with segregated stowages of special energy extracts in storage batteries, for subsequent actuation of

thousands of hydraulic and pneumatic pumps, with motors attached; 62,000 miles of capillaries; millions of warning signals, railroad and conveyor systems; crushers and cranes (of which the arms are magnificent twenty-three jointed affairs with self-surfacing and lubricating systems), and a universally distributed telephone system needing no service for 70 years if well managed; the whole extraordinarily complex mechanism guided with exquisite precision from a turret in which are located telescopic and microscopic self-registering and recording range finders, a spectroscope, etcetera, the turret control being closely allied with an air conditioning intake-and-exhaust, and a main fuel intake.

Within the few cubic inches housing the turret mechanism, there is room, also, for two sound-wave and sound-direction-finder recording diaphragms, a filing and instant reference system, and an expertly devised analytical laboratory large enough not only to contain minute records over every last and continual event of up to 70 years' experience, or more, but to extend, by computation and abstract fabrication, this experience with relative accuracy into all corners of the observed Universe. There is also a forecasting and tactical plotting department for the reduction of future possibilities and probabilities to generally successful specific choice.

Finally, the whole structure is not only directly and simply mobile on land and in water, but indirectly and by exquisite precision of complexity, mobile in air.

Tragically for a lot of people the experience of living in their bodies is a painful one, because of physical and emotional trauma, and they have retreated into their minds as a coping mechanism, rationalizing as a way of avoiding feeling. This is an effective avoidance strategy but does not really deal with the problem. The solution lies with paying attention to the pain, which will lead you towards a resolution.

Do you live in your body or in your mind? Have you left your body and haven't returned yet? Where is your 'I' located? Traditional education, the English public school model, was the quintessential process of alienation of the mind from the body and the spirit, the 'stiff upper lip' training for sacrifice and service in order to run the Empire, taking the children of expatriated servants of the Crown and the Establishment and turning them into replicas of their fathers and their fathers' fathers.

ARE YOU YOUR MIND?

The dualist philosophy separated mind from body, taking identity into the equation (as thinking), and discounting the body. 'I think, therefore I am' was Rene Descartes' assertion, but although it still speaks powerfully to us there is something missing. We are embodied (as creatures), and all the sensory information we gather comes directly through our bodies; so where we are with our bodies affects our perceptions. The world looks different depending on our position.

Descartes' ideas served to separate the realm of the body from that of the spirit/mind. In his time the Church governed the pursuit of knowledge. In order to release science

'You say to me: "I see you sitting there." And all you see is a little of my pink face and hands and my shoes and my clothing, and you can't see me, which is entirely the thinking, abstract, metaphysical me'

Bucky Fuller

from the rule of the Church Descartes managed to establish this dualism – the distinction of mind and body. The Church continued to supervise thinking related to the spirit and the mind, while science was able to explore the truths about the body.

William Blake (1757–1827), the painter, poet and mystic, pointed to the error of leaving out the body from the equation of human life and the identification of the body as 'bad', and the mind and soul as 'good'.

> All Bibles or sacred codes have been the causes of the following errors: One that man has two real existing principles. viz, a body and a soul. Two, that energy, called 'evil' is alone from the body, and that reason, called 'good', is alone from the soul. Three, that God will torment man in eternity for following his energies.
>
> But the following contraries are true: One, that man has no body distinct from his soul. For that which is called body is a portion of soul discerned by the five senses, the chief inlets of soul in this age. Two, energy is the only life and is from the body, and reason is the outward bound or circumference of energy. Three, energy is eternal delight.
>
> William Blake, *The Marriage of Heaven and Hell*

Modern thinking is bringing the mind and body together again. New discoveries show the body–mind to be an integrated, interdependent and intertwined network. 'Mind doesn't dominate body, it becomes body; body and mind are one. I see the flow of information throughout the whole organism as evidence that the body is the actual outward manifestation in physical space of the mind.' (Candace Pert, *Molecules of Emotion*)

This concept is backed up by hard science and Pert's discovery of neuropeptides. Nerves secrete substances called neuropeptides, which float around until they find appropriate receptors for their messages, which include emotional feelings. The number of receptors can increase or decrease – consequently, we can become more or less sensitive to grief, pain, or pleasure. Glands, like the pancreas, and the brain, also produce neuropeptides, and the brain itself secretes hormones like insulin. What were formerly thought of as separate nervous, hormonal and immune systems are more like one system, interconnected through the peptide communicators.

THE THREE DERMS

Biologically human life begins as a single fertilized ovum. After only a week or so of rapidly developing cell structure, two distinct layers appear in the embryonic disc – the endoderm and the ectoderm. In the third week the mesoderm appears between them.

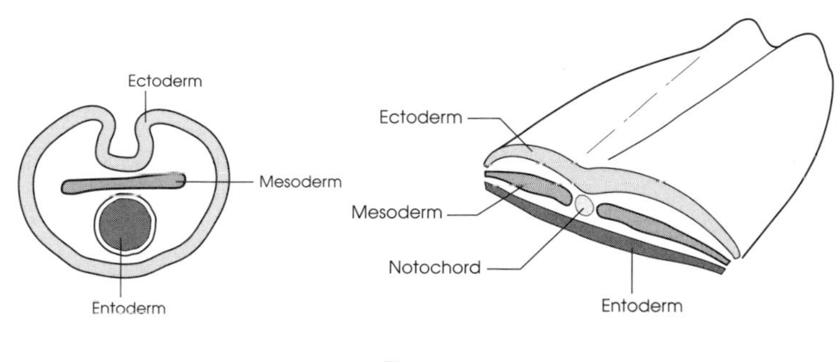

Derms

These germinal cell layers increasingly differentiate, each layer responsible for a major chunk of the developing organism. Ectoderm is the outer layer and becomes the skin and nervous system – communications. Endoderm is the inner layer – guts, energy processing and generation. Mesoderm in the middle becomes the framework and structure. In the new cosmology of the body, the skin is the surface of the brain. The skin and brain develop from the same primary tissue layer in the foetus – the ectoderm. 'The organization of our neural pathways seems to proceed from outside in, rather than inside out' (Dean Juhan, *Jobs Body*).

The endoderm is the inner germinal layer, becoming responsible for converting air, water and food into the kind of energy the body needs for survival and well-being. The endoderm also appears to have its own brain. Studies have recently shown that the 'enteric' brain, connected to the central nervous system by the vagus nerve, has 100 million neurons – more than the spinal cord and far more than the connecting vagus nerve can cope with (it has a mere 10,000 neurons). Nearly every substance that helps run and control the brain has also turned up in the gut.

Mesoderm, the middle germinal layer, develops into all the connective tissues of the body, the structural elements: muscles, fascia, bones, blood, cartilage, tendons, ligaments, and also the kidneys, gonads, heart and dentine of the teeth. This is the continuous network of myo-fascia which is vitally affected by the way we use our bodies and is the focus of the following chapters. Connective tissue, in varying degrees, is the ocean within us and contain the same basic chemical proportions and carbon compounds that are found in sea water.

The body is more a story than a machine.

The physical and mental equipment together still do not complete the picture:

> Man is not alone the physical machine he appears to be. He is not merely the food he consumes, the water he drinks or the air he breathes. His physical processing is only an automated aspect of a total human experience which transcends the physical. As a knot in a series of spliced ropes of manila, cotton, nylon, etc., may be progressively slipped through all the material changes of thickness and texture along the length yet remain in identifiable pattern configuration, so man is an abstract pattern integrity which is sustained through all the physical changes and processing.
>
> Bucky Fuller

We are the Universe experiencing itself, loving itself, through body-ness just as the rest of Nature embodies the same story.

If you can change your mind, why can't you change your body? And for that matter, if you can become wise with age, why isn't there a somatic equivalent? Is there a flaw in human structure that prevents us from being comfortable and means we become tired whilst sitting around doing virtually nothing? You and I can think of a far-away star and instantaneously 'be' there in thought, hundreds of light years away. Our bodies plod around, lagging behind

'The concentration of chemicals in our blood and tissues are precisely the same as those that once prevailed in that embryonic sea'

Lyall Watson, *The Water Planet*

our minds, never fully representing who we are, but rather where we have come from and where we have been – a fascinating embodiment of the past: the accidents, traumas and scrapes that we have survived, and the memories, thoughts and beliefs that we espoused. Getting up-to-date is a process of letting go of that past.

I believe that human beings are rather well designed. I am not going to attempt to improve upon it, but rather to reveal it more clearly. All of us are similar from the outside; most of us have all the standard issue parts: arms, legs, heads, sensory apparatus, internal organs. On the other hand, it is said that if you could see the world I see through my experience you would think you were on another planet. So there is a great deal of distinguishing difference in the way each of us holds ourselves, both literally and figuratively, in the world. The *way* each of us does what we do, as much as what we actually do, feeds back into our structure. So we must maintain adaptability in the face of the rapid acceleration of the rate of change. We can no longer afford to be sedentary.

You may believe that there is some expert out there in the world who knows more about you, or a part of you, than even you do. This is not true. This is part of the problem of alienation. Your decisions will be more effective when you take back responsibility for your own well-being. Take advice, talk to experts, but also acknowledge your own expertise. Nobody else knows what it is like to live in there. It is a Universe which only you have experienced.

All the anthropological and biological evidence supports the theory that species become extinct through overspecialization. There is so much specialization already in our society. What is missing in this world of specialization is the ability to pick up all the pieces and make meaning of the whole. You have to be in your body to know that you are not your body.

The alienation of the mind of man from his own vehicle, his body, is reflected in the state of the relationship of man with his environment generally. As we are out of touch with our digestive processes, our guts, so the wastes of society are polluting the oceans and clogging the arteries of the Earth's regenerative systems. By focusing our intellect on our own internal systems, we can discover a perspective which might just lead to more effective actions out there in the environment.

> When the whole body and the whole being is included in the educational process, the rate and depth at which learning can occur is truly remarkable.
>
> Joseph Heller

> Healing the Planet, one body at a time.
>
> Stuart Bell

You and I are more than our complement of body chemistry, physiology, neurology, reflexes, and our thinking and feeling are not the whole story. When we die everything that we appear to have been dissipates into dust and ashes – life leaving its shell behind. Many experiments have shown no change in bodyweight at the moment of death. Whatever animates us, it is weightless, metaphysical.

EXERCISES

Some may find it easier to answer the opposite question to the one which opened this chapter: who are you not? This exercise is designed to help uncover the essential you by releasing attachment.

Disidentification Exercise

Sit quietly alone for a few minutes, and let your mind ponder on the following thoughts.

I have a body, but I am not my body. My body may be sick or well, tired or rested, but that has nothing to do with my self, my real 'I'. My body is my precious instrument of experience and of action in the outer world, but it is only an instrument. I treat it well; I seek to keep it in good health, but it is not myself. I have a body, but I am not my body.

I have emotions, but I am not my emotions. My emotions are many, contradictory and changing. Yet I always remain I, my self, whether in joy or in pain, whether calm or annoyed, whether hopeful or despairing. Since I can observe, understand and label my emotions, and then increasingly dominate, direct and utilize them, it is evident that they are not myself. I have emotions, but I am not my emotions.

I have an intellect, but I am not my intellect. It is more or less developed and active. It is my tool for knowing both the outer world and my inner world, but it is not myself. I have an intellect, but I am not my intellect.

I am a centre of pure Self-consciousness. I am a centre of Will, capable of mastering and directing my intellect, my physical body, my emotions and all my psychological processes. I am the constant and unchanging Self.

SUMMARY

- You are more than just your body.
- You are more than just your mind.
- Your body–mind is an integrated, interdependent and intertwined network.
- The nervous, hormonal and immune systems are all interconnected and communicate via the biochemical mechanism of peptides.
- You are a magnificent contribution.

- ‘You and I possess, within ourselves
  and at all times,
  under all circumstances
  the power to transform
  the quality of our lives’

  Werner Erhard
- You are the most complex system thus discovered in Universe, apart from Universe itself.

Chapter 4

Air Born – Breathing and Awareness

The young world ... is intuitively sceptical of the older world's customary ways of coping. That doesn't mean the young don't like their elders. It doesn't mean they disrespect all humanity born before them, but they realise intuitively that humanity is emerging from a womb of ignorance.

Bucky Fuller, *House and Garden* May 1972, page 202

THE FIRST BREATH

Breathing, like getting hungry, is an automated function. Most of us cannot voluntarily stop breathing for more than about three minutes, much less time than you can go without food, or water. Breathing is the most vital source of energy we have, and it is the first thing we do when we are born and are physically separated from our mother with the cutting of the umbilical cord, which plugged us in to her life support system.

The significance of the first breath is enormous, physically and metaphysically. That 'traditional' (and obsolete) slap on the back, gasp and cry, with the umbilical cord being cut immediately, is being replaced by a gentler, more peaceful introduction to life. Breathing is stimulated by contact with the air as the new life leaves the fluid environment of the amniotic sac. If the umbilical cord is not cut immediately, a breathing pattern can be established on a non-emergency basis first, and vital nutrition continues to flow into the neonate from the placenta for about 15 minutes, or until the cord stops pulsating. If born into water, with no air contact, a baby doesn't start breathing until he or she surfaces.

> We are not confronting the child with anything new. We are simply prolonging the uterine condition so beneficial to development. Living in water is totally natural for a newborn. He has never done anything else.
>
> Igor Tjarkovsky

Tjarkovsky, the radical Russian water birth pioneer, has babies swimming soon after their birth. He theorizes about the innate psychic abilities of humans, controlled by delicate structures in the brain, which he likens to egg yolk. Being born into water protects these delicate structures from damage, just as you can crack an egg into water and not break the yolk.

> If the child is also born underwater, it is given an extra start in life. [Tjarkovsky] claims that sudden exposure at birth to the force of gravity after months of weightlessness in the womb, and a

> sudden, huge dose of oxygen taken in with the first breath, are two elements which affect the most sensitive brain functions.
>
> Erik Sidenbladh, *Water Babies*

Many people born into the 'civilized' world of the middle of the 20th century were born into the bright lights, harsh words, fearful thoughts and cold, unnatural surfaces of a hospital environment. The century of global wars started with the discovery of germs.

'Spiritus' means breath, and there is a vital link between inspiration and breathing. Breathing animates and moves the body; inspiration moves and animates the being. As breathing is semi-automatic, we could say that we are being breathed, rather than we are breathing. Breathing is part of the human design – we didn't choose it, it chose us.

The first breath sets a tone. It establishes us with an independent energy source and its quality, together with what is going on around us in terms of the thoughts, feelings and emotions of the mother and birth attendants, imprints on us powerfully. Rebirthing therapy recognizes this, and through conscious, connected breathing can enable people to review those first feelings and decisions made at the time.

> We learn to breathe at birth; that lesson is often poorly handled, and with tragic results. A lifetime of continuous breathing all begins with, and is conditioned by, the first breath out of the womb … The unique quality of breath, that it can be both

> consciously and unconsciously directed, makes it a link between the conscious and unconscious aspects of our being ... While expression happens in the extremities, feeling happens in the torso and is continuously influenced by the breath. The inhale relates to will. It is the embodiment of intention, drive, desire, wanting and receiving. To breathe only into the upper chest is to avoid the strong, creative energies of the lower abdomen and the sexual organs. To breathe only into the belly is to avoid the equally strong creative energies of the heart and throat ... Partial exhaling reflects a fundamental distrust towards life, or that there is not enough, and a need to stay in control, a forceful exhale further contracts the breather's energy, and may reflect a belief that we are filled with 'bad' energies, pains, thoughts and feelings and if we work hard enough, we can expel/purge them from our system.
>
> Michael Sky, *Breathing – expanding your power and energy*

DON'T FORGET TO BREATHE

Although we cannot choose not to breathe, we can and do inhibit breathing to varying degrees, through muscular stress, poor habitual movement and posture/attitude. The quality of our aliveness can be compromised by the vicious circle of not breathing freely, which leads to a build-up of more tension, in turn leading to further inhibition of the breath.

There are many ways not to breathe freely: shallow breathing, where the inhibition is extensive and movement of the ribcage and diaphragm is minimal; belly breathing, when the diaphragm moves but the ribcage does not, so the abdomen has to be pushed out (just like a baby breathes); reverse breathing, when the person sucks the stomach in on inhalation and pushes it out on exhalation; breath holding, as in sleep apnoea – a condition when the person awakens from sleep because he or she is not breathing and, as a result, does not get enough deep sleep (*see Chapter 5*).

Just as some women admit to buying shoes a size too small (*see Chapter 6*), some cleavage-conscious women squeeze themselves into push-up bras which are also too small, hoping for that extra lift which will make the difference. As a result doctors and osteopaths are now seeing cases of shortness of breath and digestive problems, such as irritable bowel syndrome and intractable constipation. Push-up bras work by being even tighter than the regular kind, themselves likely to restrict the natural rib expansion needed for full and easy breathing. The moral of this story is not to wear tight corsetry all the time, and make sure what you do wear fits properly.

Fritz Perls (the founder of Gestalt psychology) once said 'Fear is excitement without breath'. By releasing our breath we can transform the energy of fear into the energy of excitement, because it is the same energy. You wouldn't be afraid if there wasn't something at stake, and if there is something at stake then it is something you want – whether it is to survive a truly dangerous situation or something like public speaking, which fills most of us with paralysing fear. As you remember to breathe, that energy becomes available to you.

THE STRUCTURE OF BREATHING

The breathing apparatus is the meat (or veggie sausage) in a sandwich made up of the support structure (spine) on one side, and the manipulation devices (arms) on the other, and any problems with either will impinge on the breath.

The shoulder girdle is almost a separate sub-assembly of the skeleton – only being fixed at the sterno-clavicular joint in a bony sense and known as the appendicular skeleton – and is designed to hang freely from the spine/ribcage, the axial structure. Therefore it should rise and fall passively with the breath under normal circumstances, and not get in the way of breathing.

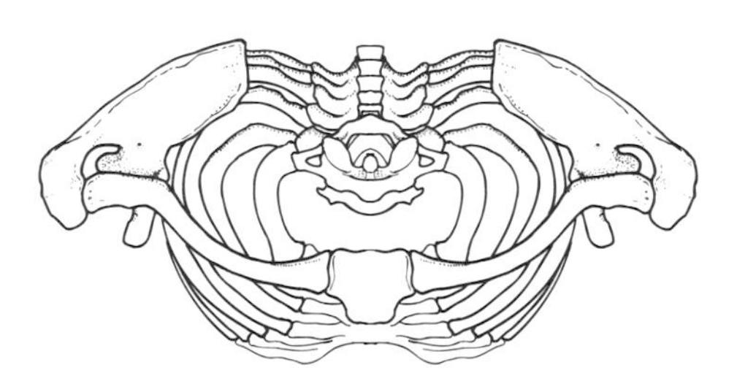

shoulder girdle from above

Similarly the spine, in an upright, vertically oriented posture, supports the head, neck and shoulders without undue tension on the supporting musculature and soft tissues, and thus frees the ribcage to expand freely rather than being burdened by having to hold up the weight of the superstructure. This part of the body can weigh up to a fifth of total bodyweight (*see Figure 2.4*). The head weighs about 7 per cent of bodyweight, on average, as do each of the shoulder/arm assemblies.

Marja Putkisto, the 'deep stretch' teacher (*see Chapter 13 on Somatic Therapies*) says that breathing correctly can improve posture, and vice versa: a neat example of two-way cause and effect in the body. This type of fact bedevils mechanical medicine and makes the holistic approach so effective. If the diaphragm moves up and down more fully and freely, it will help to organize the ribcage and torso into a more vertically oriented direction.

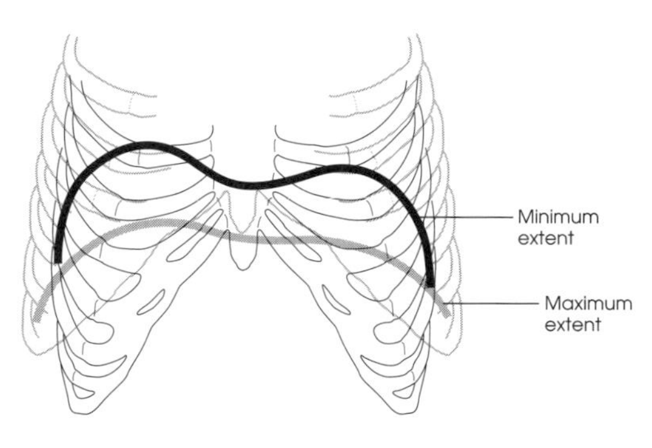

diaphragm and ribcage

Many meditation practices focus on the breath as a way of strengthening the constitution – prana yoga – and to recover past memory/decisions – rebirthing. The connection is clear: more breathing = more aliveness.

ATTITUDE

Attitude is a word which has applications for both mind/spirit and body. In ballet, 'attitude' relates to body posture; and one's mental attitude, opinion or mode of thinking also shows up in the body. If you are depressed, what could be more natural than to sink down in your posture, expressing it physically? Your attitude can become unconsciously embodied so that, for instance, after the depression lifts mentally, the depressed posture still lingers. Eventually you can't fully express inspiration even if you want to – your posture has set like a blancmange in a mould and your body has got stuck in the past.

There is a third use of 'attitude' in aeronautics. Here the word describes the relationship of the aircraft to the direction it is flying in – whether the nose is up or down, the wings level or tilted or the fuselage parallel or otherwise to the direction of flight – i.e., the angular relationship between the aircraft's axis and the wind. To fly higher you have to adopt a positive attitude, but you have to stay within certain limits otherwise you will stall! A negative attitude will take you into a dive.

AWARENESS

Awareness starts with otherness to be aware of – which itself begins with the first breath and disconnection from the mother. Separation from the wholeness of unitary consciousness causes the pain which brings us into life. Little shocks of pain wake us up to where we are, bring us to life on Earth after birth, and are part of the process of incarnation.

'The simplest descriptions are those expressed by only one word. The one word which alone describes the experience 'life' is 'awareness'. Awareness requires an otherness of which the observer can be aware'

Bucky Fuller

There are different kinds of pain, even 'good' pain, pain which teaches, pain which is feedback. How do you know when something is not right? When it doesn't 'feel' right. When, perhaps, you are too absorbed in getting a result to pay attention to subtle messages from your system, what happens then? A more insistent message starts up, perhaps an ache or a dull pain. Still not interested? How about an acute pain, headache or spasm? Is your system trying to tell you something? Are you listening? Is anyone at home?

Do you spend a lot of time getting on with it and telling your body to put up or shut up? Are you led by your brain, brain-dominant? What would it be like to have a whole-system identity, reconciling mind, body and spirit, paying attention to all the parts of you? It is your relationship with yourself that we are talking about here, before we get on to relating to anyone else, or the planet.

Awareness is the beginning of change. You cannot change something you are unaware of. It can be difficult to become aware of something like the way that you have become accustomed to breathe, or sit, stand or walk. It is so much a part of you that it doesn't register as capable of change. This is where it may be useful to get some assistance. Pain can often be the catalyst that impels us to try something new, necessity being the mother of invention. 'How bad does it have to get?' can be an interesting question. 'How good are you willing to have it be?' may lead you even further. When a somatic (body) educator directs your attention (by direct touch in movement coaching, or by guiding you verbally), to a part of you which has hitherto been outside your awareness, you can start to reclaim that part of yourself and recover your integrity. The root of the word integrity is the Latin 'tegere', to touch. Integrity can therefore be defined as being in touch with yourself.

THE AIR WE BREATHE

With a startlingly counter-intuitive theory and process, the Buteyko Method of breathing has helped literally hundreds of thousands of sufferers from asthma, emphysema, allergy, bronchitis sufferers and others in Russia, Australia and the UK to breathe more easily, by counselling them to breathe more shallowly! According to scientific fact it is essential that carbon dioxide is present for oxygen to be absorbed across the membrane of the lungs. Our lungs are never fully emptied, and the residual air in them is maintained at a constant, higher-than-atmospheric level of 6.5 per cent CO_2. Deep breathing flushes out the CO_2 and leads to an inability to absorb oxygen properly, the source of panic in asthma attacks. By controlling the breath, relief from attacks is gained. Professor Konstantin Buteyko in fact links up to 200 diseases to over-breathing.

The current atmospheric level of carbon dioxide has dropped over time to a level of 325ppm in 1982. Oxygen levels in the atmosphere have also dropped over the millennia, from 35 per cent down to 21 per cent of the air volume (a decrease of 40 per cent). Several oxygen therapies, such as the ingestion of dilute hydrogen peroxide (H_2O_2), the inhalation of pure oxygen, the use Hyperbaric chambers (which increase air pressure to force oxygen into the tissues and blood), and the inhalation of ozone (O_3), are being utilized to increase oxygen levels. Professor Buteyko says, however, that the air we breathe contains 200 times less carbon dioxide than we need, and 10 times more oxygen!

I don't want to add to the burden of dogma about breathing. My point is that if we can breathe more easily and effortlessly into all the places into which we are designed to breathe, then we can either breathe the same amount of air with less effort, or more air with the same effort.

EXERCISES

Think of the three specific places into which breath flows:

1) Your sides. The bellows effect of expanding the lungs sideways into your ribs mobilizes your rib junctions, which allows lateral flexion of the spine, which in turn frees your spine – the least accessible type of breathing for most people.

   Lying on the floor, imagine you are under a sheet of glass pressing you into the floor, notice you can still breathe sideways.

2) Your chest. This, after all, is where your lungs are, but it is surprising how few people fill the upper body with air. This is something vital to release shoulder tension – you can't relax without breathing, and especially the out breath – Aaaah!
3) Your belly. In healthy breathing, the ribcage expands its circumference on inhalation, especially at its bottom margin, so that the diaphragm does not have to piston down too far and displace too much viscera. The diaphragm does however move gently and expand

the abdomen. This area is generally easier to access and is often over-used – breathing into your belly is apparent but not actual, the diaphragm muscle marking the lower margin of the lung area.

Now breathe firstly into your sides, then, when those are full, let your chest rise. Finally let your belly expand. Then let all your breath empty and wait for the next breath. If you feel a little dizzy, slow it down and take it a bit easier. Breathing is like the ultimate exercise – repetitive movement washing through the body from the inside, a wave which defines our aliveness.

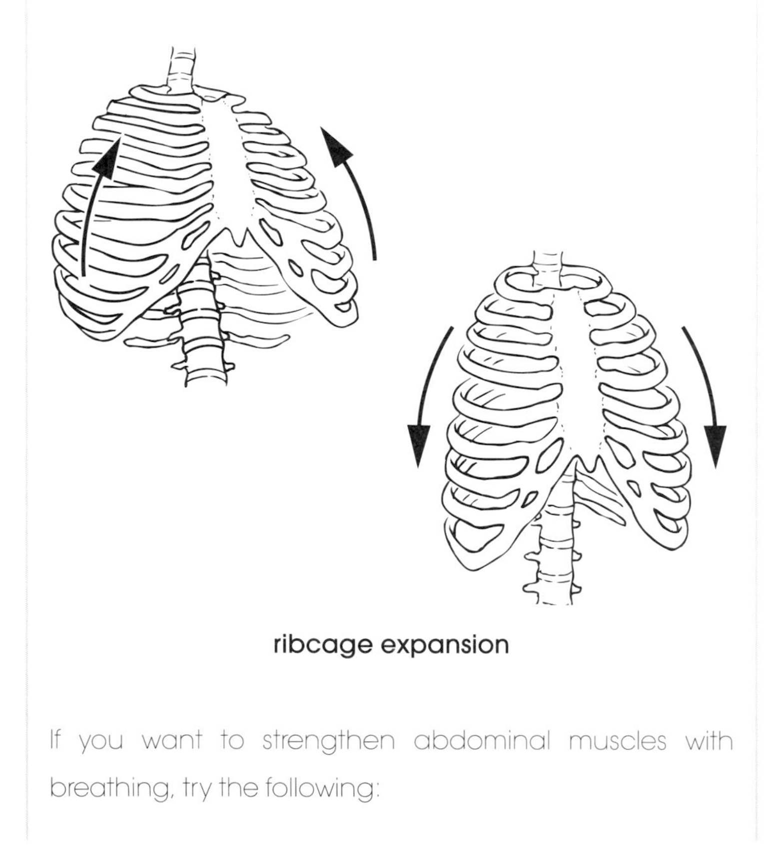

ribcage expansion

If you want to strengthen abdominal muscles with breathing, try the following:

- Sit on the floor on your heels and rest your buttocks on your heels.
- Sit up straight with your hands on your hips and inhale deeply.
- Blow every last bit of air out of your lungs, contract your abdominal muscles, and hold for five seconds ... repeat.

SUMMARY

- Breathing correctly improves posture; correct posture improves breathing.
- How good are you willing to have it be?
- How bad does it have to get?
- Giving up smoking (if you haven't already done so) is the best thing you can do for your health.
- Breathing correctly improves posture, correct posture improves breathing.
- 'Fear is excitement without breath' (Fritz Perls).
- Don't forget to breathe!

Chapter 5

Vertical is to Live, Horizontal is to Sleep

You can lie in bed
You can lay in bed
You can die in bed
You can pray in bed

You can live in bed
You can laugh in bed
Give your heart in bed
Or break your heart in half a sec.

You can tease in bed
You can please in bed
You can squeeze in bed
You can freeze in bed

You can drop in bed
You can roll in bed
Find yourself in bed
Lose yourself in bed

You can laze in bed
You can wait in bed
But never never never never never never never never
never never never can you sing in bed!
But never never never never never never never never
never never never can you sing in bed!

'The Bed' from the musical *Hair*, by permission,
© EMI, 1968

To get the really laid back sound for the Beatles' song 'Revolution 1', John Lennon ended up lying on his back on the studio floor to record the vocal track

Your mother may have been in this position when you were born, you could yet die in it. You may spend up to a third of your life in it, it is the horizontal. Some people live in bed, others retire to their beds; Hugh Hefner, founder of *Playboy* magazine made his career in and out of bed; John Lennon and Yoko Ono famously gave press conferences from bed.

LYING DOWN

Being horizontal will palliate most backaches; with the knees raised it is the rest position of Alexander Technique and, legs bent and supported at right angles, the position you are in during a rocket launch.

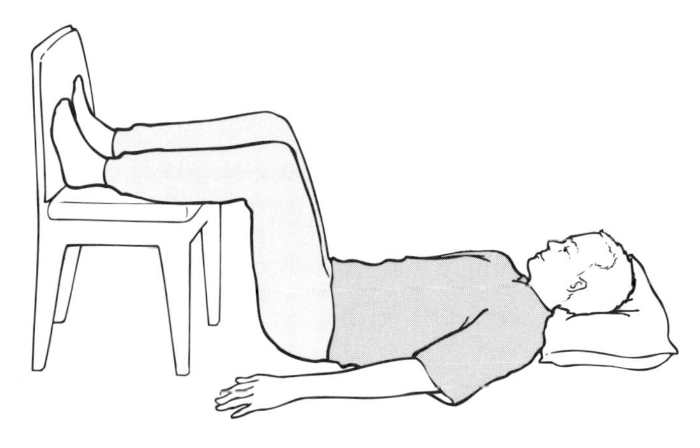

rocket launch position

This is the most relaxing position you can be in and takes no skill or effort whatsoever, which was pretty much the attitude of the American jet test pilots towards astronauts when the US space programme started in the 1960s. They described the first astronauts as 'spam in a can', decrying the lack of skill needed in that nevertheless courageous occupation. The first astronauts even lobbied for a joystick to make it look as if they were doing at least something, when in fact everything was pre-planned, computer-operated and required no interventions from them whatsoever!

More recently, the Russian cosmonauts who have spent considerable time in orbit with the Mir space station have had to be stretchered off the spacecraft when they return to Earth. Even with rigorous exercise programmes in space they still lose bone and muscle mass, and changes occur to their blood pressure and the effectiveness of their hearts as they come to accommodate a weightless environment. It all goes to show how important gravity is in the maintenance of the mesoderm and endoderm. It has proved impossible to stand up straight in space. Even staying in bed on Earth for long periods starts to have similar effects, and bed rest is largely discredited now as a medical therapy for most conditions. In fact epilepsy, asthma, heart attacks, strokes and paralysis are most likely to occur towards the end of sleep, around 5 a.m. Many people have discovered that they are better off seeing an osteopath, structural bodyworker or chiropractor than staying in bed for backache.

How you get out of bed can also be injurious to your health:

> Human beings are the only animals which will bend over immediately upon rising from bed. I can't count the number of people I have talked to that say all they did was bend over and BOOM, there was pain. Watch your dog or cat when they wake up ... the first thing they do is drop their lower backs down and extend their spines upward. You should do the same.
>
> Dr. Mark Smith, Chiropractor

UPSIDE DOWN

If lying down doesn't relieve backache, then you could go all the way and hang upside down. This practice has spread all over the world since its early origins in California. Vertically challenged prospective Highway Patrol Police recruits discovered that they could add the half or one inch they needed to join the force by hanging upside down for periods over a number of days.

Someone then noticed that it also helped with back pain, and the rest is history – and a lot of different inventions to facilitate upside-downness. These range from the early so-called anti-gravity boots – in truth ankle bands with hooks on – to the more sophisticated, easier to use and safer tilting table designs, where degrees of inversion can be graduated to suit.

In yoga, being upside-down with head stands and shoulder stands has been practised for thousands of years for its therapeutic value. More recently, former acrobat Benjamin Marantz has evolved a practice of massaging clients whilst upside-down, the clients that is. During 'inversion therapy', the practitioner suspends the client using his own body, the client hanging upside down from the thighs, supported by the feet and hands of the practitioner.

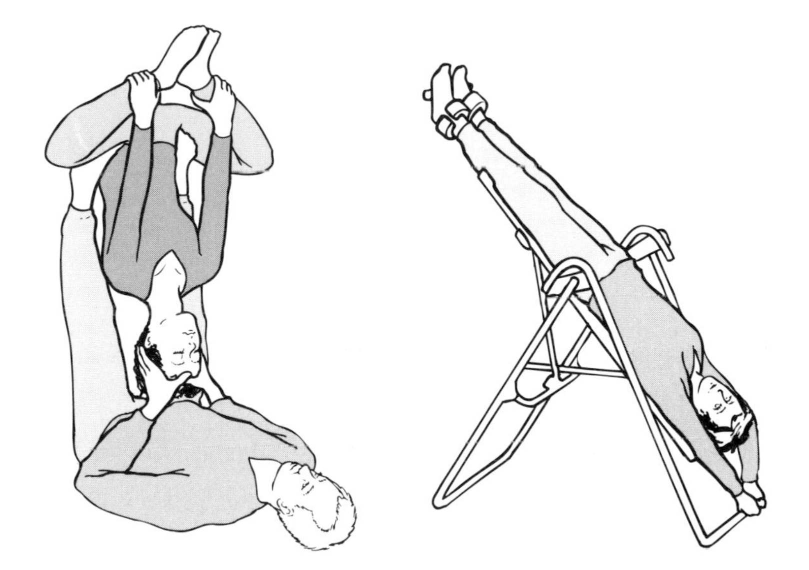

inversion therapy

Marantz's work is a cross between yoga, gymnastics and massage. Aficianados talk about the intense physicality and sublime spirituality of the experience. It is like bungee jumping without the uncontrolled adrenalin rush or the violent and forcible out-of-body experience. Other body-workers, in Sweden, have been working for years with clients hanging from mechanical upside-down tables. London trichologist Sarah Black, recommends upside-downness to stimulate blood circulation to the scalp, and thus help prevent hair loss. If you're not up to the full inversion, try hanging your head over the edge of the bed. (Other tips are rinsing your hair in cold water and head massage.)

Hanging from the ankles has its disadvantages, even dangers, unlike lying down. Although nothing in life is risk-free, staying in bed with the duvet over your head has been quoted as being particularly low in risk (apart from the danger of suffocation, that is), but don't stay there too long, as noted above – and remember to hang your head over the edge to encourage those follicles!

SLEEP

A study showed that students who got 7 hours of sleep or less per night used 3 times as many antibiotics as those who got 8 hours or more

There is no correct way to sleep. Comfort alone must be your guide, although sleeping on your stomach is not recommended for low back pain. Many people believe that they sleep on one side or the other all night, but studies have shown that the average person moves position 30 or 40 times in a night. The disadvantages of a static sleeping position become apparent in hospitals and old people's homes, where bed sores result from excessive immobility. Blood tends to pool in certain areas, and nurses have to turn patients regularly – itself a major cause of back problems amongst nurses.

In England, in the Middle Ages, and very possibly elsewhere too, it was considered quite normal to share a bed with numbers of others in a inn overnight. The Great Bed of Ware, in Hertfordshire, was 14 feet wide and was to be found in the local pub. Sleeping in the horizontal was in fact not the done thing during the Middle Ages, people were more likely to sleep, clothed and propped up by cushions in a semi-recumbent position, due to the prevalence of consumption (TB) and other respiratory illnesses.

It is painful to watch some people getting out of bed: they lever themselves up with their arms and elbows and put a lot of tension into their necks and stomachs. The really easy alternative, if you are lying on your back, is to bend your knees and, rolling onto the side of the bed that you want to get out from, swing your legs over the edge and push down onto the bed with your uppermost arm, or push your hands together and your underneath elbow down. The weight of your legs counterbalances the weight of your head, and you are soon sitting up without having expended much energy at all. If you are feeling particularly energetic of a morning, throw the covers off, bring up your knees towards your chest whilst lifting your arms above your head and, rounding your back, throw yourself, arms and legs down the bed, rolling forwards and up into a sitting position. (This movement can be continued into standing if desired, but is not recommended for those with weak hearts.)

getting out of bed

For those of you who sleep on the floor, roll onto your side and then onto all fours, before sitting back on your heels or squatting prior to standing up. Consciously push down with your feet to stand up (*see Chapter 9*).

Sleep is essential to health in human beings. For a few years Buckminster Fuller experimented with a routine of half-an-hour's sleep every six hours as an alternative to the more usual once every night pattern. We are probably designed to have two sleeps every day, like the gorilla – the siesta is a very natural phenomenon, although it is not generally recommended for insomniacs who cannot sleep during the night. Latest research suggests up to 40 per cent of insomniacs sleep far more than they realize – they just don't know that they've been asleep!

Humans sleep far more than some animals – the average

horse makes do with three or four hours a night. As a rule, adult elephants do not sleep until after midnight, and average 2 hours and 19 minutes, usually with a break in the middle. They lie down, although they can and do sleep standing up for short periods. Gorillas, on the other hand, need 13 hours and sloths and koalas even more, due partly to the lack of nutrition in their diet of leaves.

Fritz Sarasin, a Basle naturalist, made a thorough study of man's sleeping habits. Man evidently prefers to sleep with his head higher than the rest of the body: Aborigines of New Guinea use logs for pillows, others make supports out of spongy wood or fibre.

Sleep is for dreaming – deprived of dreams we can become demented, and sleep deprivation has been used as a form of torture. Dreaming, as indicated by a phenomenon known as Rapid Eye Movement (REM), has been shown to be essential to our well-being by experiments in waking people just as they begin to dream. REM accounts for about 20 per cent of our sleeptime.

A 17-inch collar size or larger, and being overweight, is a danger sign for the most serious sleep disorder: obstructed sleep apnoea, where snoring, gagging and gasping during sleep prevents the sufferer from falling into a deep REM sleep. This can lead to anxiety and depression, drowsiness and even falling asleep during the day. It is also a condition that increases the sufferer's risk of heart attack and sudden death.

Treatments for sleep apnoea include being attached all night to a machine which provides continuous positive air pressure via a mask; wearing an orthodontic device to hold the jaw forward during sleep; or even surgical removal of part of the palate at the back of the mouth.

Only tame and domesticated animals, and a few of the larger wild animals, can indulge in deep sleep without taking precautions to be constantly on guard against their enemies. According to McBride and Kritzler the *Tursiops truncatus*, a species of dolphin, sleeps only during the 30–40 seconds it surfaces to breathe. Swifts, on the other hand, have been observed to sleep on the wing, at 9,000 feet up!

'To sleep, perchance to dream'

Hamlet III,i

(Sudden Infant Death Syndrome (or 'Cot Death'), has now been related to the descent of the larynx – a normal phenomenon which occurs uniquely in humans among land mammals, and also in the dugong, walrus and sealion among sea mammals. At between four and six months of age, when cot deaths tend to occur, the larynx of the baby loses its secure contact with the palate, prior to its descent at 18 months. This is when lying a baby on its front is hazardous.)

THERAPY

A lot of therapy takes place lying down: classic Freudian psychoanalysis being a famous example, with the therapist seated behind and out of sight, and the patient supine on a couch or chaise longue, freely associating words cued by the therapist. It was Wilhelm Reich, one of Freud's early students, who realized that suppressed emotions are always stored as chronic muscular tension, which he called 'armouring', and he broke away from classic psychoanalysis to work with the body as well as the mind. This schism was the beginning of the evolution of a part of many modern somatic therapies. Classic Reichian, neo-Reichian, Bioenergetic and other such therapy is body-oriented psychotherapy where the objective is to access emotional content – going via the body to the mind. It was Dr Ida P. Rolf who developed the modern interest in human structure, and who stands as another major source of somatic therapies.

Massage is almost universally administered to a supine or prone client, with one or two interesting exceptions; some bodyworkers now specialize in working on patients whilst they are upside down (as described earlier), and there has been a flurry of interest and activity in seated massage. This practice has the advantage, to employers, that it can be administered fully clothed, 'in-house' in offices, so that stressed workers fail to get away from their desks for even a short while.

In the practice of structural integration – Hellerwork and Rolfing – treatment takes place in various positions including the sidelying position, which is effective for lower back pain and for pregnant clients and those who cannot lie on their fronts or backs. Standing footwork and seated backwork are also routinely used in sessions as part of the standard programmes offered. When horizontal, it is as if gravity is turned off insofar as the normal stress on the body is concerned, which is why we sleep reclining in this way, as it reduces the demand on the heart. Working on clients when they are standing or sitting, however, gives a more realistic picture of how their bodies are coping with the stress of gravity.

Lying down is synonymous with relaxing. But some people are so stressed that they are more tired on waking than when they went to bed, which goes to show that doing nothing can be more tiring than doing something if your body is chronically tense enough. Lying, of course, also means not telling the truth, and is associated with not being straight or upright. People who lie about are lazy. Adolescents are often thought of as particularly lazy; in fact the huge changes they are going through physically, emotionally and psychologically does cause them actually to need more sleep – honestly!

CONSTRUCTIVE REST

Lying on your back, with knees bent, on a firm surface (padded but not sagging), can be more therapeutic than bed-rest especially in the case of lower back pain. A rug, yoga mat or blanket are suitable, with a headrest if needed – preferably also firm. Feeling the ground underneath your back gives you feedback! The firm support directs your attention to areas of tension, so bringing awareness to those places and allowing you to relax and let them down under the force of gravity.

If you relax completely your body should gradually flatten down onto the surface that you are on, depending on how much chronic tension is stored in your body and whether there is any restriction in the joints from arthritis or calcification.

Can you get fit while lying in bed? It may seem an unlikely idea, but elite non-Kenyan and non-Ethiopian athletes have been queuing up to buy high-altitude bed chambers to sleep in, and mimic the reduced air pressure of living at altitude. The thinner air at altitude stimulates the production of oxygen-carrying red blood cells so that, back at 'sea-level', they can get more oxygen to the muscles and subsequently perform better on the track. These high-altitude chambers are especially popular with cyclists, where the oxygen demand is high and forms the ceiling to their performances. Miguel Indurain, the five-times Tour de France champion between 1991 and 1995, who has been described as a huge pair of lungs on legs, probably doesn't need this long, narrow cylindrical chamber for his shuteye, and as they are only a metre or so (30 inches) in diameter it is just as well.

EXERCISES

Now here is some very good news for lazy people like you and me – a couple of exercises that you can do lying on the floor! The first is a version of the rest position, and the second is known as constructive rest, where you are using the power of the mind to influence your body, just as mental rehearsal can be very powerful prior to activities like sports and performance arts. Visualizing the 'lines of action' can give the direction to your body that you need to improve your internal structure.

Rest Position Exercise

Lie on your back on a firm, padded surface with your knees bent and arms folded across your chest (see chapter opening picture), one forearm resting on the other. Become aware of your breathing, exaggerate the exhale – even 'hiss' a little eventually. Stay in this position for 10 minutes once or twice a day, especially if you are feeling stressed or tense. To come out of the position, roll onto your side and stay there for about

20–30 seconds, then roll onto all fours for another 20–30 seconds, before sitting back on your heels or knees.

Constructive Rest Exercise

Relax in the rest position for 2–3 minutes (as above), before visualizing the eight 'lines of action' below. The first four are the major lines of action, the others are secondary. If you haven't got time to do all eight, do the first four. As you become familiar with them you can do the sequence more quickly. Aim to spend about 15 minutes going through them twice a week if you can.

1) Imagine a heavy steel chain, link by link, lying on the ground under your back. Watch the chain slowly get pulled down and out under your legs until it is straight.
2) Imagine a cord coming down from the ceiling, under your knees and back up to the same place on the ceiling. Visualize the cord tightening, pulling your knees up slightly.
3) Imagine that your buttocks are two lumps of dough on an old-fashioned kitchen table. Watch the dough slowly and slightly flatten as they sit there in the warm afternoon sunshine.
4) Imagine that inside your chest is a balloon. Watch the balloon slowly deflate.
5) Imagine a wide elastic band stretched between the greater trochanters of the thighs. Watch the elastic pull the trochanters together.

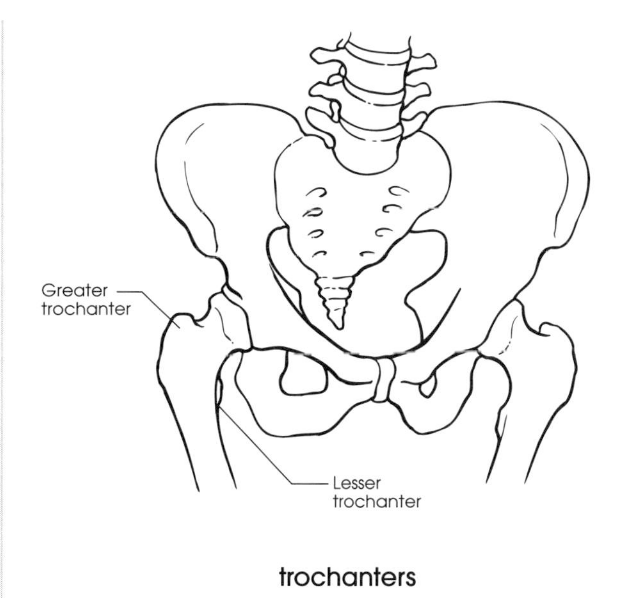

trochanters

6) Imagine that your big toes are the heads of two tortoises, watch them bring their heads back into their bodies – your feet.
7) Imagine that your chest is an egg and your pelvis an eggcup. Watch the egg falling into the eggcup.
8) Imagine that your sternum or breastbone is a ruler, watch the ruler move slowly up towards your head, parallel to the floor.

Spend some time on each image, either visualizing them in your body – in situ so to speak – or, if that is difficult, imagine them in space, whichever works for you. Don't try to make your body move, just work on the most concrete visualization you can and let your body do what it does. This is an ideokinetic process: your mind will move your body.

SUMMARY

- Lying down is the most relaxing position you can be in.
- Lying down will help most backaches.
- Getting up can strain your stomach, shoulders and neck – always, while lying on your side, put your feet over the side of the bed and use your arms to lever yourself into a sitting position.
- Enough sleep is essential for both physical and mental health.
- Therapy often takes place on a client who is horizontal – they are relaxed and gravity's pull is minimized.
- Early to bed, early to rise…
- Sleep deprivation is a very good way of torturing yourself, or others.

Chapter 6

Feet Don't Fail Me Now

Feet are extraordinarily important as the contact points between you and the planet. The quality of that contact will partly determine how much effort and energy you need to do something as innocuous as stand around.

In our 'civilized' societies a number of problems arise with regard to our feet, due to the kinds of flooring and footwear that have become commonplace.

PUTTING YOUR FOOT DOWN

The unyielding, cold, flat, unresponsive hardness of most surfaces that city dwellers take for granted – concrete floors and paving, endlessly polished smooth shopping mall surfaces, etc. – is one end of the problem spectrum. At the other is the ultra 'comfortable' cushioned training shoes that are so popular with people of all ages, and sought-after fashion objects for the young. We are covering our feet with thicker and thicker protection in order to walk on harder and smoother surfaces. In this way we are losing contact with the ground. Walking barefoot in the forest, we'd pay attention to the placement of our feet and be forced to adapt to uneven ground, rocks and roots, and so to move the 26 bones and 32 joints that comprise each of our feet.

FOOT STRUCTURE

The construction of the foot can be understood by looking at the three main contact points with the ground: two on the 'ball' of the foot, and the third on the heel. Add to this the ankle joint and the three arches: two longitudinal and one transverse, and you have a picture of the essential structure of the foot: a tetrahedronal design with springs between the contact points which give cushioning, shock absorption and self-levelling suspension.

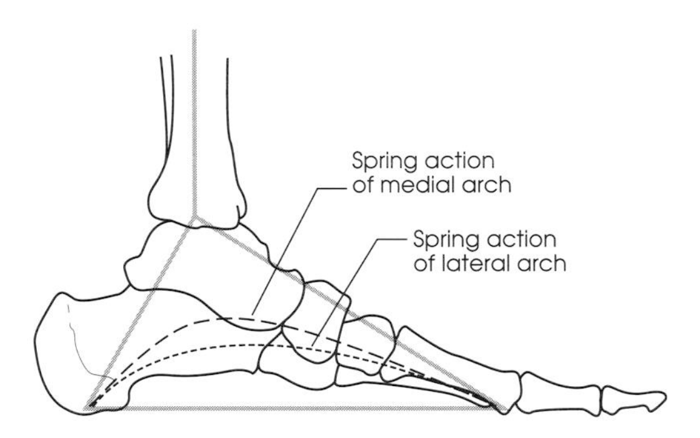

construction of the foot

'Triangle is structure, if you want structure you have to have triangles'

Bucky Fuller

I knew a ballet dancer whose arches were quite flat until he went to classes. By working at it he was able to recreate his arches sufficently to perform as a soloist for several respected companies. Every baby is born with flat feet.

Arches develop in response to the demands placed upon them. As a child begins to push up onto his or her toes, either before or after he or she begins to walk, the arches begin to form. Flat feet are not a serious impediment to walking, although it was considered a good excuse to avoid National Service at one time!

Eversion of the feet, where the feet turn out, can be a cause of flat footedness, with more weight being put on the medial arch so that it collapses under the strain. This is generally more caused by the muscles which laterally rotate the hip, as the ankle and knee joints cannot rotate laterally by themselves.

SHOE TORTURE

How and why did heels come into use in shoes and boots? When and where did it all start? What is the point of a heel, what purpose does it serve? There is some controversy about this, but my favourite story starts with the invention of stirrups for horseriding in feudal times. In order to use stirrups and stop the foot sliding through and out of position a heel was needed as a stop, and the lord of the manor instructed the cobbler to put heels on his boots. Everyone looked up to the lord, literally, so when he teetered on his new heels to get from the boot room to his horse, everyone was very impressed, and those who could afford it bought boots (and later shoes) with heels, even though they didn't have horses to ride. They became a fashion item, suggesting the existence of a steed, and hence wealth and influence in society.

Some say that platforms, a shoe style that comes around from time to time, originated in China to keep the dainty feet of ladies out of the muddy puddles that were frequently found on the streets. These are the same dainty feet which were bound with bandages from an early age in the effort to create the 'perfect' foot. The story of this barbaric practice is sobering. According to Alan Raddon, shoemaker and reflexologist:

> Small feet were much admired in Chinese society. The desire was to create the cultural rarity of 'Golden Lotus', feet measuring just three inches in length. From the age of three, the daughters' four toes were bent back underneath the arch leaving the big toe 'free', then bandaged and forced into shoes one size smaller to restrict their growth. Despite their grossly misshapen appearance, Lotus feet were seen as the most erotic part of a woman's body and were unwrapped during foreplay.

Footbinding persisted after being proscribed in 1911 by the New Republic, and was again banned by Mao Tse Tung in 1949.

Enslavement to fashion still echoes in today's shoes in the high street. We are so mean to our feet, not just squashing our feet into pointed shoes, which change the axis of the big toe and deforms the others, but in a survey 88 per cent of women admitted to buying shoes one size too small for them.

MORE SHOE TORTURE

Many women in particular feel they are under pressure to wear 'fashion' footwear. Platforms, pointed toes, stilettos and all the panoply of the shoe business border on fetishism some of the time. It really is amazing how little attention there is to comfort, support, utility, functionality and practicality in shoe design.

If you need to walk or stand in your daily life, consider the impact your footwear is having on your well-being. A heel will accentuate the curves of your spine when standing, as your body desperately searches for balance from the unpromising foundation of what is effectively an inclined surface. Your ability to walk freely, let alone run, and your connection with the ground is compromised.

How refreshing it has been in recent years to see more women wearing Doc Marten shoes and boots in the backlash against fashion footwear. Not so clever, however, is the widespread use of 'trainers', especially with the laces left untied at all times. Untied laces mean that the toes have to grip more and can lead to hammer-toe, where the tendons of the toes become permanently shortened, deforming the foot. (Note: it is best to lace shoes whilst standing in them, so that the foot is held in the right position.)

Seven out of ten people with foot problems are women – the real fashion victims

Let's be clear: the only reasons for wearing high-heeled shoes must be if you are vertically challenged, or if you have been wearing them for so long that it is painful not to do so. If you are not convinced, research from Thomas Jefferson University, Davis, in California, reported in the

American Orthopedic Foot and Ankle Society procedings (February 1992) found that the balls of the feet had to absorb 57 per cent more pressure with the shoes on than the heels, when 45 women were measured for foot pressure as they walked, first barefoot and then in shoes with just a 1-inch heel. These women were high-heel veterans – on average they had worn them for about nine hours a day, four days a week for 14 years.

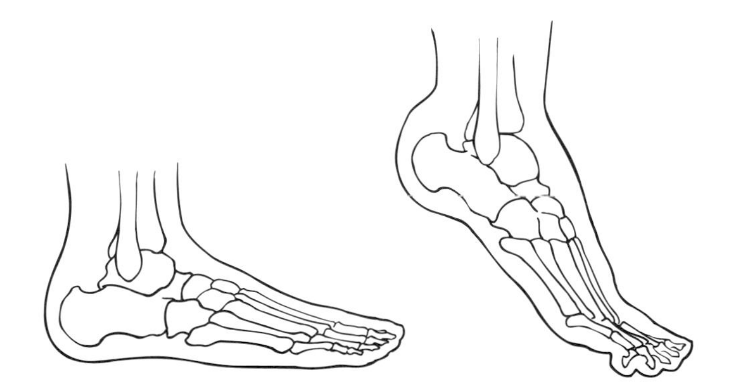

neutral foot vs high-heeled foot

Besides inducing bunions, tiptoeing around can stress your toes, ankles, knees, lower back and even internal organs. If you *must* wear heels, do all of your walking in low-heeled shoes. The degree to which the footwear takes the foot away from the horizontal will affect what we could call 'natural' walking. In general, if you want to find more ease and balance in your walking, choose the lowest heels you can get away with.

FOOT WEAR

Think of the soles of your feet as the tyres of your car. All sorts of problems arise when the tyres are incorrectly tracked or balanced: uneven wear, instability, vibration at high speeds, poor fuel economy, poor handling and road noise. The same thing can happen with your feet. You can hear some people coming from their heavy footfalls, others are tiptoeing about.

Look at the pattern of wear on your shoes. A typical pattern of wear is when the outside of the heel and the ball of the big toe wear faster. This is because the foot is tracking incorrectly on a diagonal across the plantar surface rather than along the length of the foot. If you tend to stand with the feet turned out then you may also be walking the same way, so that the outside of the heel strikes the ground first, and the inside of the ball last.

Athletic, training shoes have their place on the sports field, and have certainly contributed to improvements in performance in recent years; the actual track surface has also made a big difference. (The high-tech 'tartan' tracks have some cushioning but they are also highly responsive in the sense that they actually rebound back from pressure. The opposite of this kind of surface is another material called 'sorbothane', which absorbs pressure so efficiently that you can drop a fresh egg on to it from five feet without it cracking!)

LOOSE ANKLES

Have you ever seen a stork walk, or a chicken, even a horse? They all loosen off their ankles and let them dangle between footfalls. With our shoes and socks, and especially with boots, it is easy to forget that this is the easiest way to walk – with loose ankles. We are too busy looking out for potholes, cracks in the pavement and, worse, dogmess on our urban streets. The surprising fact is that you are less likely to trip with loose rather than stiff ankles, although you still have to look where you put your feet! I find that letting go of my ankles also helps me relax my shoulders in walking. Tight ankles go along with tight or stressed shoulders.

Loose ankles will begin to rectify tracking errors. By relaxing the ankle on the back foot in the walking action and leaving your toes behind, you are preparing to throw the foot forward on the next step. The looseness of the ankle ensures that, as you throw your foot forward, the heel is ready to strike the ground first and that the weight flow through from the back to the front is in a smooth rocking action. (Note: 'leaving your toes behind' does not mean pushing with the toes – more like dragging them, intitially, to get the idea and then later just leaving them till last.)

This action is easier in bare feet or with flattish shoes which do not need to be held on by gripping with the toes. It is harder with boots, but not impossible. The idea is to relax the ankle every time you lift your foot, and you can do that within a boot, even if the boot holds your ankle up.

The concept and practice of 'loose ankles' can be generalized and used in other activities besides walking. In cycling there is a technique known as 'ankling'. Flexing the ankle joint on each stroke of the pedal is more efficient than keeping your ankle stiff and a better use of your energy. Push your heel down at the bottom of the stroke so that your heel is lower than your toes, stretching your achilles, and at the top, leave your toes lower than your heel, so that you don't have to lift your leg so high.

Loosening off the ankles in running improves the smoothness of your action, reducing heel strike and conserving energy otherwise tied up in holding onto them. Over a long run this adds up substantially – shin splints and stress fractures are often caused by 'compartment' syndrome, where the muscles in the anterior compartment of the lower leg, between the shin bone and the fibula, become enlarged, putting pressure on the surrounding fascias. Loosening off the ankles wherever possible allows the three muscles in the compartment to relax and recover their ability to slide against one another, instead of being glued into one big lump.

Whilst swimming breaststroke it is also possible to benefit from releasing the ankles. After the leg push and glide phase, as you pull your legs towards you, notice how much more streamlined and easier it is if you release your ankles.

EXERCISES

This exercise will release the tension in the anterior compartment and in the three muscles contained within it: the tibialis anterior, the extensor digitorum longus and the extensor hallucis longus. The problem is that these three muscles tend always to work together, because of shoes and flat surfaces, which tends to permanently tense them. The solution is to mobilize them separately, or at least loosen the ankle extensor from the toe extensor muscles.

Sit, with your shoes off, on a low chair or on the floor with your back to the wall. Lift just the toes up, keeping the ball and heel of the foot on the floor. Then lift the toes and the ball up. Then (this is the tricky bit), keeping the toes up, put the ball of the foot down. Repeat three times with each foot.

As you put the ball of the foot down with the toes held up you are asking one muscle to relax whilst the others are held, allowing them to move independently. This is known as 'demi-pointe' in ballet, as in half way to being on 'pointe'. Don't try that at home!

SUMMARY

- A good overall body structure depends on good foundations – the feet.
- Feet are designed with tetrahedral structure – this provides shock absorption, and self-leveling suspension for the rest of the body.
- Modern shoes, with their high heels and pointed toes, can lead to deformed feet, bunions, pressure on the spine and a generally unbalanced structure.
- Walk in flat heels as far as possible.
- Going barefoot allows the bones and muscles in each foot to be exercised, and puts you back in touch with the ground.
- Loosening off the ankles on each step leads to more efficient walking.
- Being well grounded is the first step toward improved structure/posture.

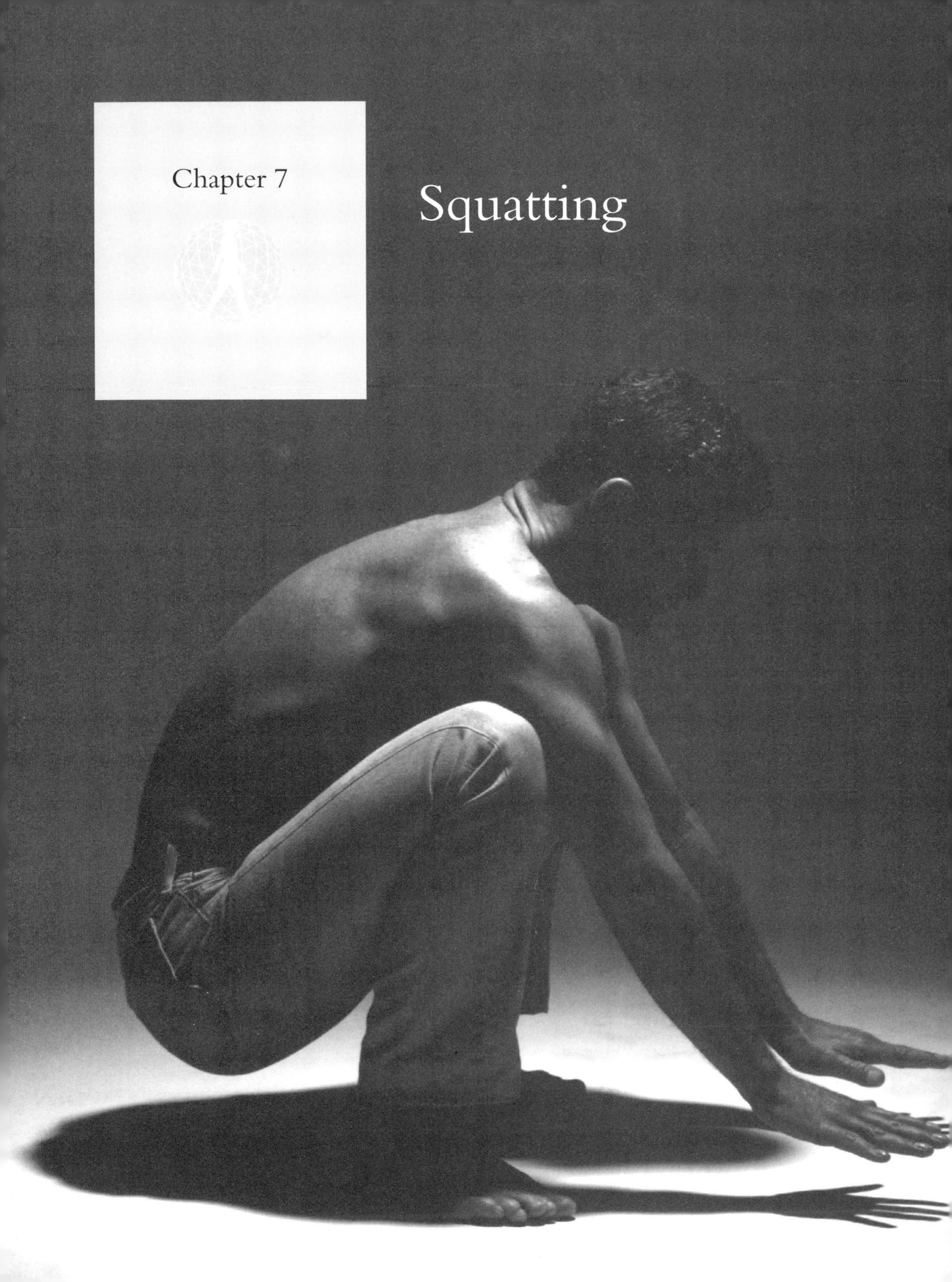

Chapter 7

Squatting

A Danish doctor once named squatting as one of the six best things you can do for a lower back. I can't remember the other five. Many people can hardly get into a squatting position these days. Human beings evolved squatting and it is still common in many societies as an alternative to sitting and, of course, as the natural alternative to sitting on Thomas Crapper's ubiquitous water closet invention.

Crapper didn't actually invent the water closet – that is a myth. He did, however, have a successful plumbing business in London from 1861 to 1904. The 'dough boys' (US troops so-called because of their oversized berets) passing through on their way to the killing fields of Flanders in the First World War, saw his name on the water tanks – giving rise to the slang 'crapper' meaning toilet.

THE MISSING HISTORY OF PLUMBING 1596–1861

The history of plumbing in modern times must begin with an honourable mention for Sir John Harington, who built a toilet for his godmother, Elizabeth I, in 1596, installing it for her use in Richmond Palace. Much ridicule resulted for Harington and his toilet, and although well used and doubtless appreciated by the Queen, there was not another like it for 200 years.

In 1778 Joseph Brahma patented an improved version. So improved was it that an original is still in use in the House of Lords, Westminster. The throne-like seat allegedly derived from the travelling commodes used by English and French monarchs, dating back to the 16th century. The technology and engineering patented by Brahma has been further improved over the years and is still in use today. Brahma's work was not taken up quickly enough, however, to prevent 20,000 Londoners dying of cholera between 1849 and 1854.

In 1859 the British Parliament had to be suspended for a short time because of the stench from the River Thames, and 1861 Queen Victoria's beloved consort, Prince Albert, died from typhoid, which led directly to the erection of a massive memorial in Hyde Park in London, where the Prince is depicted on his gold-leafed throne for all posterity. From this time on, public health became an issue of government policy.

IN THE CLOSET

> Water is so extraordinarily valuable. We don't know of any other planet with any water on it. And we flush five gallons down the toilet each time we get rid of a pint of waste.
>
> Bucky Fuller

Thankfully, since Fuller brought this situation to our attention, US federal guidelines now prohibit the use of more than 1.6 gallons per flush in new toilets installed, and incentives and tax credits are available in some states to encourage 'retro-fitting'.

Together with a diet lacking in fibre, the water closet

with its inappropriate posture has a lot to answer for. The virtual epidemic of haemorrhoids (swollen varicose veins in the anus and popularly called piles) has a lot to do with inappropriate straining to empty the bowel, because sitting does not activate rectal peristalsis in the way that squatting does. Squatting is also good for urination.

Salvador Dali was obsessed with all things rectal, and his advice is salient: never hold back if you need to go, and don't hang around waiting for anything else to happen. You can always return later as needed!

'Show me a man who reads on the loo and I'll show you a man with piles'

David Stokes, after Salvador Dali

Nobody knows what you do when you are in the little room without a telephone. This is both a problem and an opportunity – a problem because no one can coach you on improving your technique, and an opportunity because you can do whatever you want without anyone seeing you. I looked at a house once, owned by a man who had installed a French-style squatting toilet in his home in London, and it had taken him two or three years to persuade the planners that his equipment was permissible under UK building regulations. A true pioneer, although he may not have been considering the needs of some potential future occupants of the house. I, of course, was seriously impressed!

OUT OF THE CLOSET

Defecating in the squatting position has several major practical advantages. Anatomically it is the correct position to empty the bowels most easily. This position straightens out the otherwise curved lower intestine, allowing gravity to be engaged in support of this vital and virtually daily activity. In addition, the thighs support the abdominal wall, and the buttocks are spread so that faecal matter can be cut off cleanly by the anal sphincter. It is also hygenic in that you do not have to touch the seat.

Ever since I heard about the virtues of squatting I have been developing my technique as follows: first I found a wooden box or upturned, strong wastepaper basket or bucket and put my feet on it whilst sitting on the loo, thus elevating my knees to the correct position. Later, when I got fed up with having to find these accessories, I gingerly stood on the porcelain in my shoes, or on the seat in bare feet, and lowered myself into a squat. There are one or two commercial foot-stand devices around for just such a purpose.

Nowadays I can't imagine sitting at all.

My hope for the future is that squatting will be rediscovered. And with the growing awareness that water is too valuable to flush down the toilet, that we can recycle our waste products quite easily and odourlessly with low-tech dry composting toilets. American heiress Abby Rockefeller has devoted the latter part of her life to bringing the advantages of dry composting into popular use. In 19th-century Japanese inns, the guests' wastes were valued so highly as fertilizer that the rates went down with each additional person staying in the room! Every day about three billion gallons of drinkable water is flushed away, not to mention thousands of tons of useful fertilizer. All that is needed is to separate the liquid and solid waste and wait a few weeks for decomposition into a friable compost ideal for the garden to be complete. Not a viable option for apartment dwellers perhaps, but good news for ecologically minded homeowners and outdoor types. The massive popularity of hiking and climbing has made waste disposal a real issue. Dry composting toilets at remote mountain huts will allow you to leave your calling card without despoiling the environment.

If you are not connected to piped sewerage systems, composting toilets are a viable economic alternative to septic tanks and are already becoming popular. The basic design is very simple. Waste is deposited down a chute from the toilet into a coffin-like, vented chamber sloped at 20 degrees; the urine drains separately and the rest of the waste is stored pending its decomposition to an innocuous earthy compost. Variations on the basic design include fan-assisted ventilation and heating elements to speed decomposition in smaller chambers.

'We package our foods coming inbound: why don't we package them going outbound? It's exactly as easy. When nature takes so much trouble to separate liquids and solids it is preposterous to put them together again'

Bucky Fuller

NATURAL BIRTH

Squatting also comes in handy with regard to giving birth. Again, we are getting in line with gravity. There are one or two variations on a classic squat: the monkey position, where one knee is on the floor, and the assisted squat, where the mother is supported from behind by her partner or birth attendant.

supported and unsupported squatting

Giving birth lying in bed is a convenience for obstetricians, said to have originated at the court of Louis XIV in France. In this position the baby's weight will press down onto the mother's spine, and interventions such as forceps and pain relief are more likely to be needed to assist the delivery – an interesting example of an occasion where lying down actually causes the problems for which one 'needs' to lie down, as none of the interventions are possible without the mother lying supine. Having a baby in a modern, Western 'civilized' society then becomes akin to suffering from a complicated and antisocial disease, which must be managed and treated with surgical gloves, facemasks and gallons of disinfectant. What a start to the new life!

Walking around, squatting, and even getting down onto all fours all will assist the baby in getting into the best position for being born – head down, engaged in the pelvic inlet. Water births are also chosen by some mothers, although you shouldn't get in the water until labour is well advanced, because it slows everything down.

My own experience, as a father attending ante-natal classes in London in 1989, was that there was a lot of fear in the minds of the 20 or so couples present. The questions everyone had were along the lines of 'what is the worst thing that can happen?' The birthing process; pro-creation, surely the most magical experience a loving committed couple can have had been industrialized in this culture to the point where it was only fear that seemed to remain.

This fear is the one thing that needs to be left out in the birth process, or better still, transformed by breath ('spiritus', remember?) into the excitement of this great adventure.

EXERCISES

Squat! Modern life doesn't afford many opportunities to squat other than as above. The virtually daily need to empty the bowels is the best chance you will get in our Western culture to be in this position, and it has many benefits, as noted above. But try squatting whilst talking to small children, or watching TV. If you can't get your heels down or feet parallel, do what you can. The more you squat the easier it will become. Remember to get up from squatting by actively pushing down into the ground with your weight over your feet, activating and lengthening the deep abdominal muscles so the back muscles are not over-used (*see Chapter 9*).

Note: There are some devices on the market to aid toilet squatting. Warning! Get on and off the toilet carefully. If you can, support some of your weight with an arm somewhere to make sure the toilet bowl is not unbalanced as you get on and off. You will develop technique with practice – I sometimes jump off to avoid unbalancing a wobbly bowl, and put a hand on the seat as I do so.

SUMMARY

- Humans evolved squatting – although in the West we seem to have lost the art.
- Squatting is one of the best things we can do for the lower back.
- Squatting, together with more fibre in our diet, can help reduce many digestive and colonic problems, such as piles.
- The natural position for giving birth is in a squat, where gravity aids the process and the baby's weight does not press down on the mother's spine – as it does in the modern supine position (and thus often necessitating outside intervention in the form of forceps and epidurals).
- Procreation is not a disease.
- Squatting is the missing link between standing and sitting.
- Squatting is natural.

Chapter 8

Homo Sedens, or Flying by the Seat of Your Pants

The average person slouches,
furniture is designed for average people,
therefore don't expect furniture to support you if you want to sit well.

It is my contention that human beings were not designed to sit as such: squat, recline? Yes. Stand, walk, run? Of course. But sit? I don't think so. I cannot see early Man sitting on a three piece stone suite with the family in his all mod. cons. cave. Sitting down is a recent invention. The King sits on his throne, ceremonially, indicating his exalted position as a god amongst men. As man has developed mastery over his environment and developed his technology, a lot of it involves sedentary 'activity': sitting. It is a clear expression of modern Man's ascent to heights previously attained only by potentates and emperors. Shows you how far we've come, able to sit on our backsides for much of our lives and still 'earn a living'.

Sitting down looks like the easiest thing to do after lying down but, in fact, in some ways it is harder than standing. Sitting 'up' is synonomous with a more active position, and one that is definitely a challenge for most of us. The problem of sitting is that it is itself a transition between lying and standing, and then there are the technical problems of getting in to and out of sitting.

Sitting is a transition in that the hip joints are flexed or folded, and this is not an easy position to stay in – hence slouching, which in reality is an attempt by the body to get into a more comfortable state: lying down! The problem is that slouching does not quite make it! Many innovative chair designs try to solve this problem of comfortable sitting. The standard chair in your mind's eye owes more to aesthetics and economics than to an appreciation of the support needed to maintain balance and alignment. Stackability is favoured over suitability.

'Whoever invented the chair had no idea what an instrument of torture it would become for back-pain sufferers. Suddenly the business of bending your body into a right angle ... becomes ludicrous'

Caroline McGhie, *Daily Telegraph*, 10 May 1998

PASSIVE SITTING

If I had had a penny each time people have told me that they find it easier, more natural and comfortable to slouch rather than sit up, I'd have earned £4.00 by now. My mother was right; why didn't I get a proper job or become a doctor? The reason slouching is more comfortable is that we have trained ourselves to do it and the body has adapted and accomodated itself to support that habit. In general, slouchers' fronts have shortened and their backs have lengthened.

typical slouch posture

Sitting badly, a position most of us end up in at least some of the time (either through poor ergonomics, in that we are doing something in a seated position which does not match the activity, ignorance, or lousy chair design), is a very tiring condition, because we are asking our bodies to remain virtually motionless without full support or proper alignment. This is also the beginning point of a lot of repetitive strain injuries (RSI). Sedentary work requiring fine motor control is affecting the working population in a number of different occupations.

Picture those poor people whose job it is to get up very early and sit on a sofa in front of the nation, pretending to be natural, comfortable, even cosy. Yes, we're talking breakfast television – it's a tough job but someone's got to do it. The reason I mention it is that it is a great example of conflicted sitting. Your bottom half is slouching terribly because the seat is too low, soft and tilted back, and your top half is trying to look good. I'll bet most presenters have mid-back pain, where the slouch (flexion) turns into uprightness (extension).

Consider a typical 'relaxing', passive, couch-potato-type sitting position, as if you are watching those poor people on TV. The pelvis is tilted back, and the chest sinking, with the result that the head is falling off the neck. It is only hydraulic pressure, the contents of your stomach and the limitations of the skeletal and muscular structure, that prevents your body from folding up in a heap and rolling onto the ground!

The typical driving position is very similar. No wonder many people find this activity particularly ennervating and likely to trigger aches and pains. Many vehicles don't have enough headroom for tall people or those with long backs. It is only the car's shock absorbers that stop injuries being sustained through driving in this most undesirable position.

According to several surveys, men who spend more than half their working time behind the wheel are up to three times more likely to develop acute back trouble than those who don't; truck drivers have five times the risk. Two-thirds of drivers covering more than 100,000 miles a year have clinical back pain. Of those with back pain already, nine out of ten say sitting incorrectly makes their pain worse, while one in five say operating the clutch and gear stick increases problems. Back injuries from all causes have more than doubled since 1980 (Christine Doyle, *Daily Telegraph*, 3 October 1997; figures from the National Back Pain Association).

typical driving position

The classic hands-on-the-steering-wheel position we are all taught of '10 to 2' is great for racing cars, but in the days of power steering, cruise control and automatic transmission, you are better off with your hands at '20 to 4', so that you can relax your shoulders and arms, perhaps even supporting your wrists in your lap.

Compare your sitting position when driving a tractor to driving a team of horses, rowing a boat to running an outboard motor, or sitting at a computer compared to writing a ledger at a lectern. All these modern activities turn an opportunity into a potential crisis!

The best model for active sitting is probably found by looking at horseriding. If a horserider sat like the standard car driver he would be in trouble very fast, because the spine would not be able to absorb and distribute the up and down forces of trotting. In a vertical orientation the spine supports the head and shoulders and its normal lumbar, thoracic and cervical curves combine and balance like a spring.

Dr. A.C. Mandal of the Finsen Institute in Copenhagen did the original research which led to the first ergonomic chair designed by Peter Opsvik of Norway, the balans variable. The sitting position on the balans variable (the original 'kneeling stool' of 1979) encourages a concave lower back, well-balanced posture and downward sloping thighs, just like a good horseriding position. Opsvik continued to design original chairs over the following 20 years.

Looking critically at chair design and benchmarking to Opsvik's achievements in design, it soon becomes clear that virtually nothing in production comes close in terms of ergonomics. There is an additional dynamic quality to Opsvik's designs: his chairs are nearly all designed to adjust, and many are on different kinds of rockers to allow two, three or four different positions.

The market for chairs is dominated by criteria unrelated to their prime function: to support the human structure while sitting. If you think about it, it is really quite incredible that such a ubiquitous object is so far removed from the actuality of its function, leading one to the conclusion that either designers and others do not know that there are clear principles involved in sitting well, or that they are enslaved by what has gone before, or that they wilfully avoid issues regarding functionality. (This is very similar to the shoe industry – *see Chapter 6*.)

Regardless of the answer to this conundrum, responsibility still lies with the individual in this regard, because even a well-designed chair can be abused, and no chair will make you sit any way other than the way you choose. After all, a chair is just a chair, whilst we are kissing cousins to angels, made in the image of God.

At the end of the day, however, when you are tired, it is very tempting to fit yourself into whatever furniture happens to be lying around, and so, at least temporarily, you become somewhat more like a slug than an angel. There is nothing wrong with relaxing and reclining – I am all for it wherever the opportunity arises – but I am against collapsing and declining. I would not like to give the impression that you have to be bolt upright at all times, but when reclining, keep yourself from collapsing with strategically placed cushions or pillows. It is all a question of being appropriate to the needs of the moment.

Although the human spine does have a unique capacity to bend and twist, and even undulate, for most purposes we are better off using the spine as a unit, especially if bearing any weight or when driving, lifting and reclining. Reclining in an unsupported way is the cause of a lot of back pain and discomfort in travelling – especially in cars and aeroplanes – because of the averageness built into much design.

ACTIVE SITTING

If we look at active sitting – by which I mean sitting at a computer, desk or dining table – then it is very useful to have the feet grounded, rather than dangling. A footrest or even the side-bracing of the chair is fine – the important thing is to have three points of contact with the ground: the feet and backside giving a triangulated pattern. Some contact is better than none, so having just the toes and balls of your feet on the ground is acceptable, which is useful if your legs are either too long or too short for the seat height (you may be folding your legs up underneath you if they are longer or only just reaching the ground if they are shorter). The best option, if possible, would be to change the height of the seat.

Working up from the ground, the next important element of active sitting is a level pelvis. The French word for pelvis is 'basin', and you can think of your pelvis as a bowl, the rim of which being roughly equivalent to your waistband. A level pelvis is one where, if your 'basin' was full of water, there would be no spillage. Additionally, in this position, you begin to notice the sitting bones of the pelvis, the ischial tuberosities, upon which you can balance your bony parts and internal structure.

This is where tight jeans become an encumbrance, unless they have a high Lycra content and hence stretchability. The flexibility of the hip joints determines the ease and comfort with which the pelvis can be horizontalized – together with muscles surrounding and stabilizing the pelvis (like the hamstrings and rectus femoris in the leg and the rectus abdominus and obliques in the gut). Tight trousers stop you articulating your hips.

Consider Opsvik's 'variable' kneeling stool, now 20 years old, and much copied in cheaper versions sold on special offers in their thousands. It is not the fact that you are kneeling and sitting on a forward tilting surface, as much as the way the knees are lower than the hips that makes this extraordinary chair so comfortable. I keep one in my office because just looking at it communicates the essential principle of comfortable active sitting: *KNEES LOWER THAN PELVIS*. It also challenges the mind with an intersting question: 'Is it a chair, and if so, how does it work, and why is it unlike any other chair?' I don't actually recommend it for long periods of sitting for a couple of reasons. First, there is no back support, and second, knees and shins can become sore, and having your feet on the ground is generally better than kneeling for any longer than half-an-hour or so. Opsvik himself has subsequently designed other chairs which address these very issues; the 'actulum' being one I like.

Actulum

The key angle is that of the leg to the torso. The hips are more comfortable for most people when this angle is more than 90 degrees, and in the region of 110 degrees for comfortable sitting. Wedge-shaped cushions placed on conventional chairs assist this openness in the hip joints, as does that annoying and childish habit of rocking forwards onto the front two legs of the chair when at a desk! It works! Try it.

This explains why it is so hard for Western adults to sit cross legged on the floor. Because our hip joints are not super flexible, our knees stick up and the angle of the leg to the torso is less than 90 degrees. Try sitting on your heels/knees instead, it is much easier to keep the back straight and maintain equilibrium. By raising your backside with a cushion or telephone directory, sitting cross legged can be made easier. A dense, malleable cushion of the right size can be jammed between the heels to make sitting that way easier too, taking pressure off the ankles. I find that keeping my shoes on makes it a lot easier to sit on the heels.

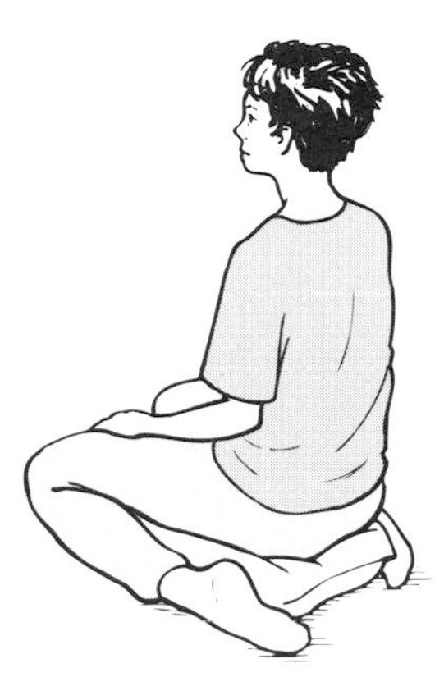

alternative to sitting cross-legged

In the spirit of unreasonableness the other best strategy when you find yourself in an unfriendly chair is to do your best to ignore it. Slide forward towards the front edge of it and level up your pelvis, or, if it is too low, raise yourself with cushions. Chair too high? Try elevating your foot height with telephone directories or other similar solid objects.

Now you've got ground/foot contact and a level pelvis we can move up to the torso. Part of the advantage of a level pelvis is that the abdominal contents can sit in the 'basin' rather than spilling out forwards, and the spine can grow out of the back of the bowl without having to contort. Imagine a tree growing on level ground and a tree growing on a steep bank. The one on the steep bank has to twist and bend to find the light.

The level pelvis, together with the vertically oriented

spine, gives maximum support for the ribcage, neck, shoulders and head. This allows easier breathing, more relaxed arm and hand movements and less tension in the neck and shoulders while holding up that heavy sack of potatoes – your head. (The average head weighs 10–15lbs (4.5–7kgs), remember?)

The height and angle of the table or desk, together with keyboard, VDU and/or the book or papers you are looking at, are all going to make differing demands on your upper body position. If you are bent over a computer on your lap on the train, it will not take long for stress and strain to appear. If your desk is flat, or your screen too low and too far back, you are going to have to bend forwards and down to be able to see what you are doing.

So as well as your legs and pelvis, you must look critically at the three-dimensional layout of your workstation. You may be able to bring what you are doing up and towards you, rather than sending yourself down and forwards to it. This, of course, is another great opportunity to be unreasonable and fit the circumstances to your needs rather than vice versa!

You can and must become creative here. You are now the expert on ergonomics because it is *your* body that is being accomodated (or not) and you are on the spot! I wouldn't recommend putting your soup on a slope – this situation is one where the old adage of 'bringing the food up to the mouth' works better than sending your mouth down to the soup. (I am sometimes unorthodox enough to hold my plate in my hand rather than leaving it on the table, although this may be too much for polite company!)

A writing slope is, however, a very effective low-tech

'In the world of office design, the battery chicken approach is applied by most employers with myopic zeal. Human beings may come in different heights and weights, but they must all occupy the same space, and sit at the same chairs at desks of the same height. When their backs (or shoulders or necks) "go", then they must hobble away into a corner and get themselves seen to'

Caroline McGhie, *Daily Telegraph*, 10 May 1998

solution to problems encountered when reading and writing non-electronic text (books). Harking back a little to the 'old days' when standing at a lectern was commonplace, the writing slope is an ideal partner for a seat with a wedge-shaped cushion. Both the chair and the table are then changed to incline towards one another and the result is a more ergonomic interface between you and the furniture. The slope brings what you are doing towards you, rather than you having to move down to it, and the wedge creates a helpful seat angle.

good office ergonomics

The correct height for a desk is so that your elbows can touch the surface when you are sitting at it without hunching your shoulders or having to bend your back. If you put your hands on the desk the forearms should be horizontal and make a 90-degree angle with the upper arms at the elbows, reducing muscular strain or the need to stress the wrists. The seat height will affect this positioning, the seat needing to be high enough to allow the hip-joint angle to exceed 90 degrees with the feet on the ground.

Some of the latest ergonomic workstation designs are now returning to the concept of standing at a lectern, reflecting the idea that it is easier to stand than to sit, and desks and chairs are being jettisoned in favour of flexible 'meeting spaces' where different groupings of employees, clients, customers and suppliers can interact. This reflects changes in corporate culture, with structures becoming more fluid and less hierarchical. Interesting, isn't it, that we use the word 'corporate' (or 'embodied'), for business life. The 'corporate' world. Which is the real corporate world?

SUMMARY

- Sitting is a transition between lying and standing.
- Slouching is the body's attempt to get into a more comfortable, lying position.
- The best example of an active sitting position is riding a horse – the thighs slope down, the feet are supported, the pelvis is level, and your *knees are lower than your pelvis.*
- When sitting at a desk, your elbows should be able to touch the surface without your back being bent or your shoulders hunched. Your hips should be at an angle greater than 90 degrees to your torso, and your feet should be on the ground.
- There is nothing wrong with reclining and relaxing, but declining and collapsing is a problem!

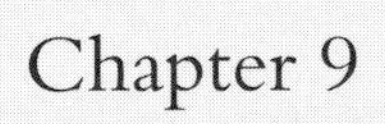

Chapter 9

Standing Up for Yourself

To every action there is an equal and opposite reaction

Newton's Third Law of Motion

Have you ever thought about how you stand up? I am talking about the process of getting from sitting to standing. There doesn't appear to be a distinction in the English language between the process of getting up from sitting to standing, and standing itself.

You probably take this movement of standing up rather for granted. It is not something that you will have really thought about. Yet it is one of the most common and frequent movements you make in daily life, so it will be instructive and possibly worthwhile to stop and consider how you get to standing up for a moment. How do you do it? I hear you saying – 'I just well … do it!' Yes, but how? This is where Newton comes in.

PUSH DOWN TO GO UP

Now there are definitely easier and harder ways to stand up. The more I think about this the more I realize that the process of standing up is a miracle: it is automated, it works virtually every time, and we do it many times each day without a thought.

The most common problem with standing up is a shortening of the muscles of the neck and back, a bracing which also occurs in sitting down. Many people also use their shoulders/arms to assist by pushing down on the arms of a chair, a desk, or onto their own thighs, levering themselves up.

Once you become clear that standing up is a process of pushing down you can begin to improve your technique in terms of effortlessness. The major part of this is to transfer your bodyweight over your feet as soon as possible, so that all the pushing can occur without the need to brace, stiffen and shorten yourself to stabilize an unstable situation.

Some call this the 'gravitational' or 'ground reaction force' (g.r.f. for short). I like to call it 'going down to go up', because there is no up without down, and generally speaking there is nothing up there to pull you up, so all that is left is to push down on something. Unless you have something to pull up from, there is no other way.

Getting your feet underneath you allows you to engage your strongest muscles (your thigh muscles) fully, and can be done one of two ways: by moving your feet or by moving your body. In practical terms either are acceptable and both together are preferred, lessening the amount of adjustment needed by either. Imagine that you are on a straight-backed chair: by sliding your feet underneath you and leaning forward slightly from the hip, you can engage your thigh muscles and press down into the ground. Your back is maintained as a unit, straight but angled forward, and your neck and head are an easy continuation of that line. Your shoulders are relaxed. You can stand up without tensing your body – you extend yourself down and get stood up by that good old Third Law of Newton. You standing up is the equal and opposite reaction to you pushing down.

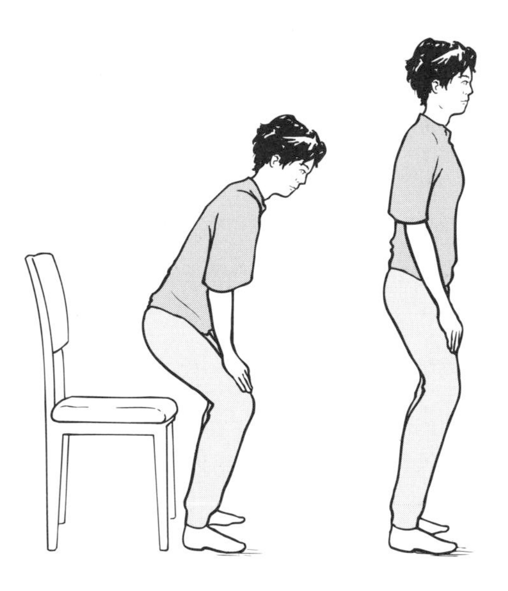

getting standing

To reiterate this in terms of the physics of the situation: you stand up when you push down because the mass of the Earth (6,600 million, million, million tons, remember?) compared with your weight is such that your push converts quickly into movement in the opposite direction. The Earth does move (albeit infinitesimally), you move more significantly, because momentum = mass × velocity.

If, instead of sitting on a hard, straight-backed chair, you are reclining on a sofa (never sit on a sofa, please, go on, surrender, recline), then you have further to travel to get into a position to stand up from; nor will you be able to get your feet back underneath you. Therefore you will have to move yourself forward over them. Move forwards until you can get to the point where you can bend from the hip and, keeping your upper body as a unit, keep going forwards until you can get the weight onto your legs and push down with them without falling back.

What we are looking for is being able to stand and sit without losing balance

What we are looking for is being able to stand and sit without losing balance, i.e., you could stop at any point in the process without effort, or without falling back or forwards. Your weight should be fully supported at all times by your feet and your ground contact.

Sitting down is a reversal of this process. Much habitual sitting consists of falling back into the chair with a crashing and compressing of the spine and neck, and the attendant moaning and groaning.

Crucial to the process of both sitting down and standing up is the ability to flex the hip joint with the leg fixed, in other words to bend the torso forwards as a unit over the legs. If you flex your hip fully you can maintain balance over the feet and do both these actions without strain.

Think what you need to do if you were to go from standing to squatting. You have to lower your bottom towards the ground, sticking it out some, and bend forwards with your torso and head at the same time to counterbalance yourself. Now put a chair in the way, and hey presto, your bottom finds the surface and you bring your torso/head to rest on top. You are now sitting. To stand up, tilt your torso/head forwards, as a unit, so far that the weight comes onto your feet and off your bottom and push down into the ground with your thighs (the quad muscles). Move your weight from your seat to your feet. You will soon be standing up for yourself.

The same principles can be applied to getting up and down from sitting on the floor. The technique is to get the weight over the feet as soon as possible and engage the thigh muscles instead of using the back (the extensor muscles). Put some weight through your hands onto the floor and untie your legs, getting the balls of the feet onto the ground. Then come back into a semi-squat and start pushing the ground to engage the thighs, and benefit from that ground reaction force!

LIFTING

To lift something other than your own bodyweight (itself an achievement when done effortlessly and elegantly), is to demand even more attention to the use of yourself and your body mechanics. The conventional wisdom and standard advice is to bend the knees, but this does not go far enough, and does not communicate the whole message. From what we have already revealed regarding your relationship with the ground it may be clear that what is needed is first, to bend the knees (yes, of course), and then to use your legs rather than your back for the lift, by pressing down rather than lifting up.

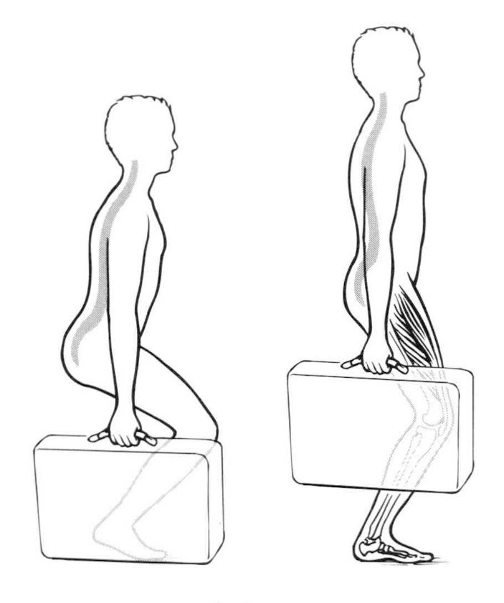

lifting

BENDING

If you are picking something light off the floor, like a pencil for instance, you can use the action as an opportunity to let go of your head and release and lengthen the muscles at the back of your neck and base of your skull. The strange paradox is that it is the very fact of keeping your back upright by holding on with the muscles that can generate back pain; the way out of it is to learn to bend freely – not in a loaded-weight lifting situation, but when you are tieing a shoelace, say, or putting a sock on.

After you have let your head drop in a no-load bend, always return upright leaving your head until last. This means you will be uncurling your back from the bottom up, using your feet and pulling your waist back.

In all bending it is far better to bend the knees and ground yourself – the same actions as discussed in lifting and in the processes of sitting down and standing up (above). As an exercise to stretch the hamstrings, touching your toes is now recognized as not safe.

Bending and lifting at the same time is also potentially dangerous – as when you try and get a suitcase out of a car boot. Lifting and twisting is even more problematic, for example putting a bag up onto a luggage rack behind you in a railway carriage without turning around fully .

SUMMARY

- There is no up without down.
- Get your thigh muscles to do the work, by getting your feet underneath you as soon as possible in the movement.
- Getting standing and getting sitting are probably the most common movements that we make. Doing them better is one of the best 'exercises' that we can 'do'.
- Let your knees move out away from you as you get sitting and standing (after Alexander).
- For lifting – bend your knees and use your legs rather than your back.

EXERCISES

Sitting Roll

1) Sit on a flat seated or forward tilted chair, with your feet flat on the ground underneath your knees. Drop your head forwards, chin towards your chest, then, relaxing your arms/shoulders and supporting your body by pressing down into the ground with your feet, transfer the weight forwards over your feet and let your back follow your head over and down towards the ground, rolling down, letting your head, shoulders and arms all dangle.

   In that position breathe smoothly for a few moments and feel the weight of your head and arms/shoulders tugging gently on your back and your back lengthen slightly on each out breath.

   To come back up, leave your head, arms and shoulders hanging and your back long, and push down with feet and thighs to the point where you engage the abdominal muscles, bringing your waist and back back. Keeping your back long and, lengthening the front to match the back, unroll progressively from bottom to top so that your head comes back up last. Check that you do not use your shoulders.

Standing Roll

2) Stand with your back against a wall, feet hip-width apart, parallel and away from the wall, so you can press your hips back into the wall. Bend your knees and drop your head forwards (as above) and roll down the wall, keeping contact with the wall until the last possible minute as you continue to roll down. Breathe slowly and gently, as above, releasing your back and allowing it to lengthen just from the weight of your head, sholders and arms. Return upright by pressing strongly down and back with the feet and thighs and unroll up the wall, pushing your front back towards the wall.

(This exercise can be done without the wall if you feel able. Be cautious if you have low blood pressure, as you may feel faint doing this.)

Reminder

Every time you go from sitting to standing or from standing to sitting, remember: press down to go up, and press down to control going down. Notice how often you forget! Becoming aware of what you do is the first step to making any desired changes, so don't be too hard on yourself for forgetting!

Chapter 10

Are you in Good Standing?

Now that you have stood up, how long can you stand it? Standing still is quite an exercise in itself, and many of us find it difficult to remain standing for even five minutes without shifting weight from foot to foot, leaning against something, or looking for somewhere to sit. Standing still for any period is a very good test of how well organized you currently are in terms of the structure of your body.

Gravity is a strict teacher. If you are not in line with gravity it will let you know quickly. On the other hand, gravity rewards you when you line up with it and it will help you to stand tall. Paradoxically, gravity, which pulls everything down to Earth, is also the force which allows you to extend upwards away from it, and it is only by using gravity, by opposing it, that standing up is possible at all. Standing up is a constant and dynamic process of extension downwards and, hence, upwards (*see Chapter 1*).

Think what it takes to stand on a surfboard riding an ocean wave, or keep your footing whilst standing on a train without hand holds. That is what I mean by dynamic and constant. Now remember that you are on a more or less spherical spaceship, spinning around on its axis and flying around the sun at over 65,000 miles per hour!

UNDERSTANDING UNDERCARRIAGE

The last thing that that surfer above could do is to lock his knees, because he is making constant adjustments with his legs, not to mention the rest of him, to ride the waves. If you are locking your knees when standing you are losing that dynamic relationship with Earth, and missing out on that equal and opposite reaction force. Standing will be more tiring because you are not allowing yourself to recieve that support. Support is available – you are not alone! You are in a relationship with the Earth.

Knee locking is usually a sign that the overall body balance is not over the legs, and that it is too far forward. Knee locking often goes together with locking the hips. Try locking your knees and wiggling your hips whilst standing – it is just not possible. Try to find the centre of balance, neither locked nor bent, so that your weight falls into the ground and is not held at knee level.

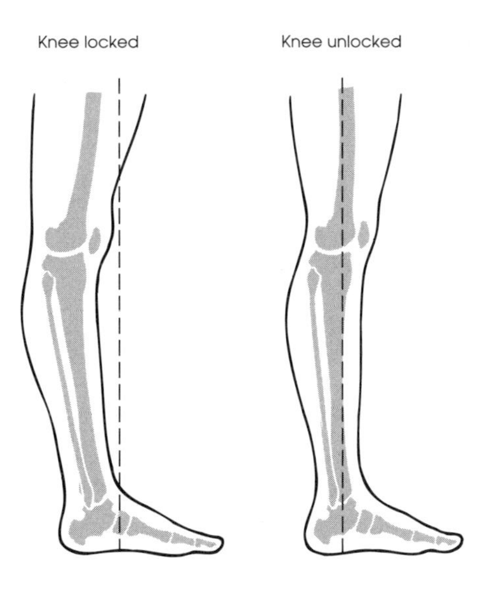

locked knees

Years of knee locking may have stretched posterior knee ligaments and hamstring insertions so that hyperextension feels 'normal'. Unfortunately it isn't; it is only 'average'. Retraining yourself can take time. If you catch yourself slipping back into locking the joints, then at least you are becoming aware that you have a choice, and that is the beginning of change. Don't expect to change old habits totally and instantly. Everything happens in its own time, but by becoming aware of the habit you have initiated a process which can continue, and energy will follow thought.

Because we are looking at getting the legs – the foundations of our structure – underneath us, let's look now at the width of stance. The Queen was asked once how she managed to stand for so long at garden parties and investitures where she met hundreds and thousands of people. She replied that she kept her feet hip-width apart.

The foundations *must* be underneath the superstructure. Too wide a stance and support is lost; the weight cannot fall through the bones and the muscles end up bearing our weight, hardening and eventually calcifying (becoming more bone-like) due to the demand put on them. A narrow stance, with legs together, will tend to be unstable and require constant small adjustments, just like walking on a tightrope.

AXIAL SUPPORT

In a balanced human standing structure we are looking for a level pelvis, the same as in sitting. An (anteriorly) tilted pelvis allows the viscera to spill out of the bowl, puts a strain on the abdominal wall and does not support the upper body. The lumbar spine is also displaced too far forward. It is as if the whole mid-section is too far forward. This is the person who pats their stomach and says they need to lose a few pounds. Often the person who slouches while sitting, with the pelvis tipped back, has the opposite style of standing. This is the commonest pattern seen.

Making sure the pelvis is horizontal is a central goal of structural integration. The pelvis is the centre of balance for the whole body, the attachment point for major muscle groups, the foundation of the spine through the sacrum and the sacro-iliac joints and the container for the reproductive, digestive and eliminative systems of the body. In Structural Integration work (e.g., Hellerwork, Rolfing, etc.) the pelvis is approached from above and below, rather than through a frontal assault, and this approach seems useful in a movement-based programme, which we are outlining here, due to the centrality and complexity of the area.

Bringing balance into the pelvis in the function of standing cannot be achieved without reference to the 'guy-ropes' which hold it in position above and below. Shortened or tight hamstrings, adductors, abdominal muscles, or lower back muscles and structures will all affect the position and orientation of the pelvis. The standard 'tuck your tail under' instruction doesn't work for this reason, and usually results in squeezing the buttocks, increasing the effort involved.

The curves of your spine are a very important feature of the human design.

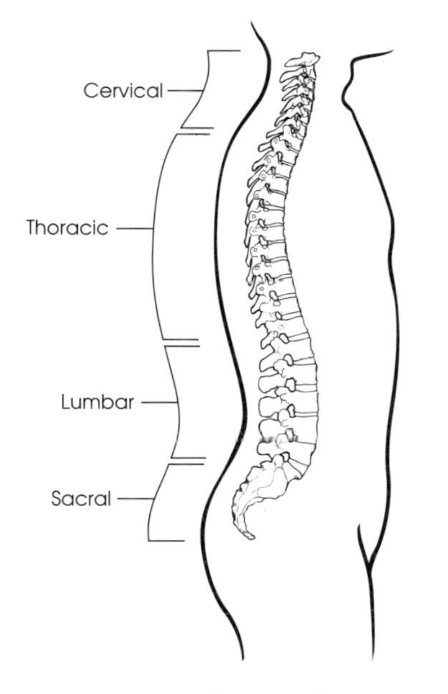

spine and curves

There are four naturally occurring curves. The thoracic and sacral curves are with you at birth. They formed with you in the womb. The cervical and lumbar curves, however, develop when you are an infant, when you begin to hold your head up, sit up and crawl. This is one of the reasons babies are better left on the floor to develop their bodies naturally, instead of being put in laid-back recliners so that they can see what's going on, before they learn to hold their heads up and sit by themselves at about six months old.

The ribcage forms a strong but moveable container for the heart and lungs. In standing, as in sitting, we want the chest on top of the pelvis, so that the weight of the upper body is evenly distributed around the rim of the bowl of the pelvis, and the diaphragm is free to do its job properly. Too far back increases compression in the low back and the ribcage tends to sag down. Too far forward and the body shortens in the upper back and neck.

THE UPPER GIRDLE

The position and condition of the ribcage also affects the shoulders and neck/head carriage because, although in the skeleton the head is balanced on top of the spine, in the reality of the living structure, many muscles attach from the neck and shoulders onto the trunk. Many people's ribcages are somewhat more 'glued' onto their spines and shoulder girdles than the design demands, so that breathing, shoulder movement, head and neck carriage and spinal mobility are all knotted together.

How many of us think or have been told that we have 'rounded shoulders'? If you are a parent, have you been telling your children to pull their shoulders back? Please don't do it. 'Rounded shoulders' are invariably falling forwards off a sinking ribcage and an anteriorly tilted pelvis. Teach your children to sit better (*see Chapter 8*) and their shoulders will straighten effortlessly. Pulling the shoulders back just creates additional stress on the back, between the shoulder blades, which will match the shortness at the front. So you end up with twice the original tension! *Gently* pull them up from the head/neck or by a few strands of hair, and get them to use chairs which are the right height and angle, so that their feet are on the ground or on an appropriate footrest.

Compare your mobility in this area to that of a cat – no contest!

The military positions of 'attention' and 'at ease' demonstrate two tiring ways of standing. In 'attention' the legs are touching (so the base is narrower than the hips), the chest is thrust out and shoulders and arms held back. While in 'at ease', the legs are too far apart to give support. The easy way to stand is to enrol gravity in supporting you and relax, without collapsing, into better alignment.

Your arms are meant to hang down freely until they are needed for action. This can only happen if the shoulders are not engaged in holding the structure up. They cannot relax until you let yourself be supported from the ground by a more 'gravi-dynamic' organization of the rest of you.

When you get your chest above and your legs below the pelvis, you can bring your pelvis into some kind of balance. This can be done visually: while standing, without moving your body below the neck, glance down to your feet. If you can't see the fronts of your ankles, try to bring your middle back until you can. From the side, look at the relationship between the outside ankle bone (lateral malleolus), the centre of the knee joint and the hip joint (greater trochanter) where it sticks out on the side. Drop an imaginary plumb-line down and see if you can connect all three up on the line. When wearing trousers, look at the side-seam and make sure it is vertical. Don't forget to pick your head up again!

So, let us be upstanding, don't let's have a stand off, or let me stand you up. Stand out from the crowd, be outstanding, and understand that you can stand up for good standing without using a stand in!

EXERCISES

Centre Up

1) While standing, notice the balance of your weight over your feet and rock slightly back and forward, front to back, to find a centred position. Relax your knees so that they are neither locked not bent – a middle position which allows the weight to fall into the ground with the least effort.

Little Plié

2) Bend your knees a little, bringing your hips back over your feet and very slowly push into the ground, maintaining the effort until you straighten up again, purely by pushing down, feeling an extension upwards right into your neck and head. This can be done almost imperceptibly whilst standing waiting for buses, trains, etc., or even at social events; or it can be done in a more exaggerated way as a formal exercise to activate

your personal gravitational reaction force (*see Chapter 9*).

Loose Shoulders

3) With a bigger and faster knee bend, go up and down so that your shoulders jiggle, but purely from the up and down body movement, and therefore passively from the point of view of the shoulders. See how freely you can allow your shoulders to bounce up and down. (Note: this can also be done on a rebounder, or whilst walking downstairs.)

SUMMARY

- How long can you stand?
- Unlock your knees in standing and bring the weight back over your feet – more on the heels.
- Conversely, don't bend the knees either – find the middle point of balance.
- You should be able to drop a mental plumbline down your side from your ear, through your shoulder, hip joint, knee joint, to the outside of your ankle bone.
- By getting your foundations underneath you properly, gravity will actually support you.

Chapter 11

Perambulation and Carrying

As sitting has become more and more common, so walking is becoming rarer. Car ownership continues to rise, the pace of life increases, who has time to walk? Yet walking remains the most complete exercise: it can be slow enough to be meditative, accessible to people of almost any standard of health and fitness, and capable of bringing aerobic benefit and releasing endorphins if done for long enough. I know a client who weaned herself off anti-depressant drugs through walking and walking until the natural endorphins kicked in.

WALKING AS A PHYSICAL MEDITATION

Some people walk on a treadmill; that's fine. I like to integrate walking into my ordinary routine. The important thing is not where you walk, but how you walk and the consciousness that you bring to it. When you walk too fast on a treadmill you haven't time to monitor your technique because you are too busy keeping up with the machine. What an opportunity missed!

Slow down a litle. As you are walking you can check your technique – loosening those ankles to release tension in the lower leg (*see Chapter 6*), relaxing your shoulders, keeping your head and your eyeline up. Walking can be a bastion of unconsciousness, but there are many variables to consider when walking: from the movement of the ankles and knees to the pelvis, back, shoulders, arms, hands and elbows, head and face. Slowing down allows you to focus on efficient and effective walking. (It is a strange fact that walking fast is inefficient, which is why it is great conventional exercise; see below, *Exercise – Walking*.)

Wherever possible, go 'hands-free' when walking, so that you can allow your arms to swing. You don't have to pump your arms or do anything with them, just allow them to move naturally and easily, and keep your gaze ahead of you rather than looking at your feet. Open your heart area as if your face is in your chest and you are looking out at the world from your heart. Keeping your head and chest up releases the shoulders to move freely.

It takes the best part of a year for an infant to learn to stand and walk. What an achievement! The first time little Johnny walked there was such celebration. Now he's fully grown, he is still walking just like he did when he was a baby. A lot of imitation goes on at an early age; you might copy your dad's walk and, even though you don't need to limp, you might because he does. Later on, as an adolescent, you saw Clint Eastwood in the movies and adapted your walk to take in a little bit of Clint. That's the way it goes: no hint of a rational thought about the mechanics of walking, or what works best – it is all imitative, and if what you are imitating is poor the chances are that you will not walk well either!

THE IMPORTANCE OF THE PSOAS MUSCLE

It is fluidity that is missing from our walking. You see this in Aboriginal peoples – a wave that flows through the body on every step. In contrast Westerners can often be inhibited, stiff, compressed, tentative, withheld, repressed or plain awkward. Walking can express the accumulation of character, the encrustations of time, the barnacles of burden. Fluidity is connection right through the whole structure.

A vital muscle, if not *the* vital muscle in walking is the psoas, a deep muscle which links the leg with the torso across the pelvis. This muscle is unknown to most people but it is the mechanism which allows the flow of movement to continue up from the legs in walking. Imagine for a moment that your legs end at the tops of your thighs, so that movement occurs only from there down. Can you see that this would produce a very 'mincing' walk? Only through allowing movement to continue upwards, across the pelvis, can the upper and lower halves of our bodies be integrated in walking.

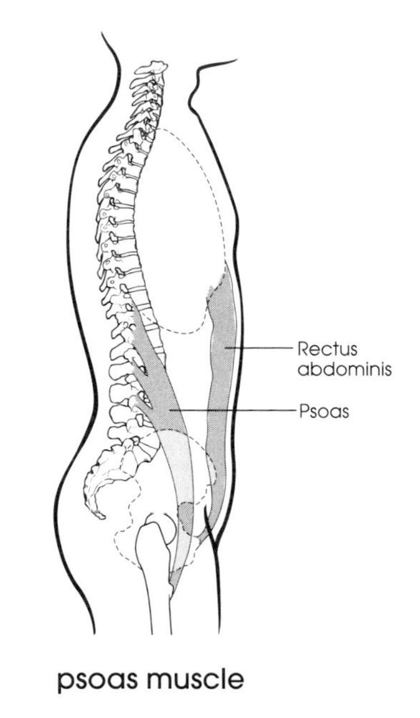

psoas muscle

The psoas muscle is the filet mignon – the principal flexor of the hip and thus central in walking up stairs (where the leg is raised when the torso is fixed) and bending from the hip (the torso flexes when while the leg is fixed). With awareness of your psoas muscle, your legs can continue up into the solar plexus region, in a muscular sense, and your walking can be more effortless, graceful and integrated.

The power of this muscle is demonstrated by a footballer or javelin thrower. The psoas muscle is engaged by the lengthening action of stretching either the leg or the torso backwards prior to the effort of the kick or the throw.

The archetypal side-to-side walk was that of Marilyn Monroe, who allegedly had one stiletto heel a half inch shorter than the other, to accentuate her side-to-side wiggle

In this new model of walking the pelvis is the fulcrum of movement between the legs and the torso; the psoas passes across the pelvis but is not attached to it. So in balanced, integrated walking the pelvic movement is forwards and back rather than from side to side.

A good image for the pelvis is that of walking with a

brimful bucket of water; you have to swing it forwards and back because if you try and hold it still it will spill. (The archetypal stiff pelvis walk is that of John Wayne.)

Imagine a window opening out onto a beautiful garden. Visualize the details: the kind of window, the kind of frame, glass, etc. Now locate the window in your abdomen and open it up, lengthening the front of you, but not by shortening your back, only by opening the front. Now remember you have long, long legs which start high up just under your diaphragm, that your pelvis swings like a bucket, and that your knees are on strings being pulled forward parallel to the ground with your ankles loose, and your foot tracking heel to toe!

As walking becomes integrated the possibility of that fluid wave becomes more explicit: an undulation that starts with the foot gradually pressing into the Earth and travels up the leg and into the spine; from the spine into the neck and then out through the head.

Imagine a heavy skipping rope spread straight along the ground between two children. One child lifts her end of the rope and quickly pulls it down again. A wave travels along the rope towards the other child. That is a picture of an undulation. Learning to undulate can take time, and, to begin with, will be tiring. You are asking for movement at a very deep level. This is a significant challenge. Practise for a couple of minutes at a time, and don't give up! Eventually the wave can go right through you to the neck and head. It is probably easiest to learn whilst lying on your side, but try it prone as well.

When you get the principle in your body you will be able to experiment doing it standing, sitting and walking.

CARRY THAT LOAD

It is when walking that we carry things, turning ourselves into beasts of burden. When we carry inefficiently it compromises our walking, increasing the load on a usually already unbalanced structure and accentuating any problems. Apart from when using a back pack, carrying also immobilizes the shoulder girdle in some way, preventing one or both arms from swinging freely and counterbalancing the leg movement.

As a rough guide, if you do not want to do serious damage to the muscles, discs and ligaments in your back by carrying uneven and excessive loads, don't carry more than 10 per cent of your own body-weight, and *use a rucksack* or bag with good thick straps over both shoulders.

unbalanced and balanced carrying

Wheels on suitcases and well-designed backpacks and child carriers can lighten the load. Specialist trekking equipment for hikers leads the way in this area and their designs will leak into the mass market as time goes on. The best packs have a high centre of gravity, padded and contoured straps and many other features not yet available in the general marketplace.

Some interesting research was done with women in Africa who carry loads on their heads. It was proved that women of certain tribes use less energy to carry substantial weights than trained army recruits using backpacks. In 1995 the journal *Nature* published a paper explaining that most people conserve up to 60 per cent of their energy during normal, unloaded walking by transferring potential to kinetic energy using the principle of a pendulum. The African women in this survey did this even better, conserving up to 80 per cent of their energy. This allowed them to carry up to 20 per cent of their body-weight without any increase in calorie consumption, and up to 70 per cent of their body-weight with less energy usage than a huffing and puffing European 'control' group. The Europeans even tried to copy the Africans by carrying lead weights fitted to bicycle helmets. 'It didn't work out' said Dr. Norman C. Heglund, 'From an energetics point of view, there was no savings. It was the same cost as using a backpack. And we managed to get stiff necks for our efforts.'

Interestingly, further research showed that part of the method used by African women involves them raising and lowering their gravity centres and their heads on every step, providing them with additional forward momentum on the up-swing of each cycle, and that this energy saving technique only works in a moderately fast walk. When the women were asked to walk more slowly they soon broke down under their heavy loads!

Growing children can be badly affected by carrying heavy and/or unbalanced weights. The widespread practice in the UK of children carrying their books between different classrooms and the absence of desks or lockers in schools is putting strain on young bodies. A survey reporting 40 per cent of 14 year olds saying that they had unaccountable back pain is very worrying. Under the British 1992 Manual Handling Regulations, no one at work may carry a load exceeding 20kg (44lbs) for more than 10m (11yds) without resting, so it is surprising that there is no similar law to protect schoolchildren in the UK. In Bolivia, Colombia, Egypt, France, Israel and Poland there are such rules; for instance, the maximum weight girls under 16 can carry in Bolivia is 5kg (11lbs), and in France the figure is 8kg (17.5lbs).

Norma Montague of The National Back Pain Association, states that, 'Excessive and uneven loads are accelerating the degenerative process on the intervertebral discs and ligaments, joint capsules, muscles, facet joints and general immature skeletal structure, increasing the risk of back trouble and permanent deformation of the spine.' A slightly long-winded way of saying it is rather bad news for children to carry heavy bags. In France every child has a back-pack. There are whole aisles of them to choose from in the shops, and it is considered standard issue. Not only that, they wear the straps over both shoulders!

The fashion of carrying a backpack on one shoulder makes a sow's ear out of a silk purse. My ideal design for a backpack will make it easier to get the pack on and off, as I think this is one reason why so few people bother to get the second strap on. Shoulder bags are bad news generally, and the heavier they are the worse for you. The worst examples have got to be golf bags and laptop computers. Help is at hand however. Fanny Sunessen, Nick Faldo's former caddy and a useful golfer herself, has invented an ergonomic golf bag strap which is being widely marketed. As a smallish woman she had to do something to alleviate the torture of carrying a full set of men's clubs. Very simply, the strap allows the weight of the clubs to be carried over both shoulders.

So called laptop computers (probably the worst place to put them) are inevitably being carried in bags with shoulder straps or as hand-held luggage. They are much too heavy for most people, and I know several who have stopped using them because of the pain that carrying them was causing. Again roll-along luggage and well-designed backpacks would be an option, until the technology moves on and portable computers are so light and ergonomically friendly that the problem is bypassed, which will not be long.

EXERCISES

Walking

If the two main objectives in exercise are to elevate the heart rate and to burn calories by using the major muscle groups in the body, then walking accomplishes both, because it becomes highly inefficient the faster we walk! (While running, your leg is like a spring, storing energy from each impact and converting it to an upward push.) Push your walk towards jogging speed to burn the most calories.

Walking makes you feel great, keeps you fit and is virtually injury-free (unlike running). All you need is a good pair of walking shoes. Choose shoes with a firm heel cup for stability, a rocker sole to enhance heel-to-toe motion, and plenty of room for your toes to spread out as they push off.

Spinal Undulation

1. Side-lying

You can practise your own undulatory movement lying on your side on a thick rug or carpet. Support your head with a forearm. Start by 'spooning' your pelvis back and forth and allow your spine, chest, neck and head to follow the movement freely. Initially the message may get lost somewhere in your mid-back. Think of your entire ribcage pivoting around an axis through your sides: as your pelvis rocks, your ribcage rocks, slightly later as the wave moves through you. Don't forget to allow your head to follow the movement too, pivoting on an axis through your ears. Keep sending the message with your pelvis, initiating the wave. Sooner or later the message will get through.

2. Prone position

Lying prone on a flat, padded surface, put your hands under your forehead to stop your nose getting squashed. Begin this movement by rocking your pelvis into and out of the surface you are on. Rock it subtly, trying not to squeeze the buttocks, up and down so that you alternately lengthen and shorten your lower back just above your sacrum. Now allow that up and down movement to initiate an up and down, in and out movement into the middle of your back. Keep sending the message from your pelvis through your spine like that skipping rope. You squeeze yourself along the surface you are on. It is like being a caterpillar going along a twig. It is how you would move if you didn't have arms or legs.

3. Camel walking

To bring undulation directly into walking is, I fear, beyond most of us except as a feeling of a possibility, a potential or an ideal. There is an exercise known as camel walking, where you send an undulation up and through the body on each step. It certainly looks comical the way I do it. If you can get this movement in your body you are in the advanced class.

SUMMARY

- Walking is the most complete exercise there is.
- When walking, keep your head up, your eyeline up and your shoulders and ankles relaxed.
- In integrated, balanced walking the movement is forwards and backwards, sending an undulating wave up through the body.
- Carrying heavy loads in an unbalanced way can seriously damage the human structure, especially if you are still growing – use a back-pack over both shoulders, rather than a heavy shoulder bag over one.
- Get hands-free in walking wherever possible.
- Loosen ankles in walking.

- Release shoulders/arms.
- Don’t forget to breathe.
- Walking is slow enough to think about *how* you are doing it.

Chapter 12

Perspiration – Running and Sports

I make a distinction between Physical Education and Physical Training. The latter represents the submission of the body to the will of an outside force, be it a set of ideals or a formula; the former is a reflection of inner vitality with some kind of self-expression involved.

At the brink of the third millennium, the pitfalls of the mechanization of sport and fitness are becoming apparent. The 'fitness' and health and leisure 'industry' (well-named by its proponents), has become a major employer. In order to deliver a marketable and profitable product to the public, this 'industry' has followed the well-worn path travelled by the medical and educational establishments and detailed by Ivan Illich in his books *Medical Nemesis – the expropriation of health*, and *Deschooling Society*.

The scenario goes like this. If you want to be healthy you must see a doctor. If you want to get educated you must go to school. And now, if you want to be fit you must join a gym. This represents an expropriation of fitness; something I see as an inalienable human attribute, yet very much unexamined, undefined, and not well understood.

FIT FOR WHAT?

A bit like 'If only I lost 7lbs, then I would be OK' is for the diet and weight-loss business, 'I must get fit' is the thought around which the whole fitness industry organizes its resources. Like other impossible-to-fulfil ideals, it takes a process and turns it into an end result, subverting it in the process. Without even knowing what the end result is, the consumer can be seduced into purchasing a stream of equipment and services, external bolt-ons which look like they are part of what is needed for the intended result, but are not what is really important. No will power? Get a personal trainer. Poor self-image? Do some weights. Feel guilty about your lousy lifestyle? Do a tough aerobics class.

The burgeoning study of Sports Science has defined fitness in terms of three factors: stamina, strength and suppleness, with a tendency to concentrate on the first two at the expense of the third. Stretching has been a poor relation, often unattended to with the time pressure on exercise routines fitted into or around busy lives.

I would like to suggest a couple more factors in the definition of fitness:

1. Structure

Fundamental to my critique of the way fitness has been appropriated by the 'industry' is the lack of awareness of appropriate human structure. There really is little point in running a marathon with a collapsed structure except to accelerate wear and tear, which one must asssume sensible people wish to avoid. As a triumph of the will over the weakness of the flesh it remains a stupendous achievement, but at what cost?

A good example of the importance of structure to the efficacy of exercise is jogging. The variation in benefit from running and or jogging must depend critically on the geometry of the running body. Jogging in fact has been largely discredited, due to the deleterious effect of repeated impacts on joints and the joggling of internal organs. Running, an altogether smoother action, a bit like the difference between a horse trotting and cantering, is perhaps less harmful but a much more technical exercise.

From an engineering point of view there are three types of stress which can affect structure. First, compression, which has things squashed together too much; second, tension, where they are stretched apart too much; and third, torsional stress, the twisting force which combines the two simpler forces.

When the structure of the body is aberrated the degree of stress it will suffer just from standing, sitting and walking rises sharply. As soon as you increase the demand on that structure through exercise, that stress skyrockets. It makes a lot of sense then, to improve or optimize structure prior to exercise.

2. Spirit

Mechanization is the key problem with the industrialization of fitness. Human beings are not designed for endlessly repetitive routines like the early 20th-century production lines. Yet exercise is defined as repetitive movement against resistance. The danger is one of alienation, of the separation of mind and body. This is what is happening when, in the latest gyms, the video screens flash and the music pumps out at high volume with a repetitive rhythm and the bass turned up. What is that doing to the person? Some find it an environment alien to themselves – perhaps they are protected from the desensitization that occurs by finding it repellent. I have also met people who think there is something wrong with them because they don't feel good in what, after all, is supposed to be a place which epitomizes health. As if it is their fault!

BRAINS OVER BRAWN

Wouldn't it be great if just thinking about lifting weights could make you stronger? The notion isn't as strange as it sounds. We already know that athletes can improve performances by mentally rehearsing beforehand. It was in part this success that inspired a group of researchers in 1991 to see whether visualizing a workout could also improve strength.

The researchers used an exercise machine to measure the leg strength of 24 women. Over a three-day period, half the women met in a quiet room for 20 minute sessions in which they were told to imagine contracting their right quadricep muscle. These 'rehearsals' were monitored to make sure they didn't 'cheat' by actually contracting the muscle. The other women neither exercised nor thought about it. When the researchers measured leg strength again, they found that when the mental exercisers actually lifted leg weights, they had increased their thigh muscle strength by 13 per cent. The control group showed no change (*Journal of Orthopaedic, Sports and Physical Therapy* 13 (5), 1991 (pp. 231–4)).

In a similar UK-based study by psychologists at Manchester Metropolitan University, one group of male students were asked to do a specific exercise every day and another group just to imagine doing it. After eight sessions, the group who actually did the exercise improved the muscle's response by 30 per cent, and the group that had imagined doing the exercise improved their muscle's response by 16 per cent.

Analysing different activities in terms of balanced body usage will be useful to anyone contemplating which one to take up as a general fitness activity, if they are not attached to any particular one. I recognize that this isn't necessarily a rational decison – it is often based on enthusiasm and the inspiration of seeing others excel at something. It also depends on how you do what you do. In swimming, for instance, breaststroke, especially with the head held up, is not recommended for back-pain sufferers. They are much better off with frontcrawl and backstroke. Think about your symmetry when you swim, or ask a friend to observe you and tell you what they see. Swimming is one sport where a little coaching can go a very long way.

BRAWN OVER BRAINS

The origins of the gym environment must be noted. The modern explosion in gym usage started when bodybuilding contests began to be televised and 'pumping iron' became a popular way of building up muscles to rival Mr Universe's. However, Charles Atlas and the early bodybuilders started as a very off-beat minority and the relationship with health is very tenuous. The main purpose of bodybuilding is to look tough so that bullies don't pick on you.

Martial arts come in many varieties and styles, some still being evolved. The tradition of the master passing on skills to students has continued for a couple of millennia. A teacher may have many students, but only one or two will be chosen for their dilligence and interest to continue the line, the rest being a bit like cannon fodder. They get what they get, but do not receive the essence of the teaching, which is integrative, holistic and spiritual.

In pre-war fascist Japanese society, martial arts were subverted by mass politics and used as a way of channelling aggression and for training a population into the highly disciplined and fanatical war machine it became that obeyed without question.

Many of the most popular martial arts are products of this process and are devoid of their original spiritual content, concentrating instead on street fighting techniques and aggression. The teachers of these 'hard' martial arts were schoolchildren in the 1930s, when everyone did these exercises and, coming to the West, found that their skills were marketable.

Other fitness regimes have developed from similar origins in the West: the Women's League of Health and Beauty being a particular product of the Empire, with uniforms, massed displays of discipline and military precision.

THE PARADOX OF THE OLYMPIAN

Some years ago I treated a very well-known athlete coming to the end of his career as a middle distance runner. From schooldays he had trained to become the best in his field, and he was. He explained to me that middle distance running at the higher levels is a controlled sprint; in other words, for the whole 800 or 1500 metres you are up on your toes. This leads to adjustments by the body to accomodate the demands which are being placed on it. In this case the Achilles tendon and the whole calf area shortens because the demand is for the body weight to be on the toes, a different kind of balance from the everyday type, where body weight is distributed over the whole foot. He noted that middle distance runners generally have injuries, strains and problems in the calves, just as the sprinters have problems with hamstrings.

The athlete in question, just from training, had lost the ability to keep his heel on the ground, so specialized had his body become, like a horse whose anatomy has evolved for running to the degree where the hoof is an adaptation of two toes and the rest, with the heel, never touch the ground. He stretched, but had lost that ability and walked around as if he was wearing high heels – on the balls of his feet.

'A runner gets a stress fracture when repetitive pounding and muscular action on the leg exceeds the strength and reparative capacity of the bone. Stress factures are found in only two other species, the racehorse and the racing greyhound, and then only when forced to run under human supervision'

Casey Myers, *Walking* (Random House)

This athlete was as graceful as a throroughbred at the peak of his career, with world records and Olympic Gold medals to his name, but towards the end his head carriage became strained and the years of training at the very edge of his capability sadly took their toll, because he had lost his ground support.

I wanted to work on his neck and shoulders – but he insisted I only attend to his calves. His mind was becoming as specialized as his body and he had lost a sense of his wholeness.

To me this is the paradox of the Olympian: that in order to excel in a specialized activity at this level, to be the absolute best in the world bar none in these competitive times, you have to dedicate yourself so fully and go to the edge of what is humanly possible, to the point where you can lose general adaptability, which is the basis for humanity's success – the ability to adapt to changing circumstances.

PROFESSIONAL PITFALLS

All professional sports require specialized skills; some sports could be considered bad for your health if you look at them objectively and take into account the demands placed on the body, in seasons that are too long and arduous with too much competition and not enough rest. Looking at the one-sidedness of tennis and golf, for instance, it is easy to see that when the professional games took off it became vital for 'cross-training' to become a part of the routine of players. Just playing the sport, as the amateurs had done, for the love of it, led to crab-like tennis players and corkscrewed golfers.

Many so called 'injuries' in professional sports are in fact akin to RSI. (They are not actually injuries at all, but, rather, occupational hazards – the inevitable result of over-training and over-use.) Golfers' backs, soccer players' 'groins' and knees, etc. are the long-term hazards of the endless repetition of, for instance, kicking a ball or swinging a golf club. Some golfers have more stressful actions than others but all are one-sided, and all run the risk of damage given their playing schedules. One-legged soccer player's – those that kick mostly or exclusively with one leg – are at risk of a separation of the pubic bones of the pelvis, at the symphysis pubis, from the constant relative strain of standing on one leg and kicking with the other. Other injuries constantly reported are damage to the ankle and knee ligaments.

Soccer players are particularly unlucky in the sense that even an extremely successful player cannot reduce the number of games he plays. He must determine when to retire in order to get substantial time off. Some very successful golfers play substantially fewer tournaments than rookies need to compete in and remain competitive, but it is not easy to find this balance in highly competitive professions like tennis.

PITFALLS OF MECHANICAL EXERCISE

Even in the gym there is trouble afoot. How many times have you heard it said that strengthening abdominal muscles will help prevent lower back pain and problems? Unfortunately scientific studies now suggest that strong 'abs' lend little or no support to the spine and its movements. Also, in order to do a proper sit up you must force your lower back into the floor in order to isolate the abdominal muscles and prevent the hip flexors from engaging. This is where problems can occur, because the normal curvature of the spine protects it from misalignment and injury. Over time the proper spinal curves can actually be lessened, straightened, or even reversed, and the probability of problems occurring increases dramatically.

Other gym-based injuries also occur. *The Times* carried a report a few months ago on the dangers of exercise bikes:

> Dr Goldstein quotes a 55 year-old patient who, after his heart attack, 'bicycled' 50,000 miles: his heart was strong but his penis limp. Where the penile artery had been compressed against the pubic bone by the saddle, it was irretrievably scarred and damaged.
>
> 'Body and Mind', 23 July 1998, p.20

Standing toe-touches and full neck rotations are also on the black list of activities, along with the classic sit-up (legs extended, hands behind neck). Recommended exercises for firming the waistline and strengthening the abdominal muscles are fast walking, swimming, cross-country skiing, push-ups, biking, surfing and trampolining (rebounding).

The burgeoning cult of Fitness could be causing other problems: exercise addiction may be more widespread than previously thought. Two recent surveys of physiotherapy clinics and aerobics classes reported in *Focus* (June 1998), suggested that up to 1 in 16 people attending showed signs of exercise dependence. Tell-tale signs of addiction include trying to exercise when injured, withdrawal symptoms such as headaches, nausea, vomiting and depression, and refusing to have rest days.

Not as easy to understand as cigarette, caffeine, sugar or alcohol addiction, where gratification is immediate (and the downside pretty obvious), this dependence may be related to the release of endorphins, the body's own opiates – the withdrawal symptoms are similar to those from opiates. Exercise addicts are often type 'A' personalities: obsessional over-achievers in professional jobs, whose behaviour puts them at risk of heart disease and other stress-related disorders.

EXERCISES

Changing the 'Same Old Way'

Feldenkrais says if you do something too fast you'll do it the same old way. Holler says that under stress you'll also go back into old habits. Try doing simple tasks, like brushing your teeth, or your hair, with your 'other' hand. Try and take your first step upstairs with your 'other' foot. More advanced versions of this exercise might include turning the car the hard way: clockwise with a left hand drive, anti-clockwise with a right hand drive. You may not have realized how 'grooved' you have become, doing things the same old way.

Here are some exercises to break habits which cause rotation of the spine (and therefore torsion in the structure):

1) Carry your bag on the 'other' shoulder or arm.
2) Learn to throw a ball or frisbee with the 'other' hand.
3) Learn to kick a ball with the 'other' foot.

SUMMARY

- The modern gym industry defines fitness by stamina, strength and suppleness. Structure and spirit should be added to this list.
- Many sports injuries are not injuries, they are occupational hazards.
- Excessive and exclusive training eventually damages the body's structure and ability to adapt.
- How do you 'fit' into your life? Maybe if you stand, stood and walked well you would be fit.
- Integration of mind and body is a valuable and vital part of fitness.

Chapter 13

Somatic Education: The New Gym

Somatic education is a learning process taught by a teacher or practitioner whose intention is to teach the individual a process that creates physiological change and that the process of change can become self-initiated and self-controlled, enabling an individual to have greater voluntary control of their physiological process. The word 'soma', refers to the body, the whole body. It is the body as perceived from within and is inclusive of all aspects of being human.

Daniel Foppes, *Towards a Definition of Somatic Education and Practices*, 1995

Aston Patterning, Pilates, Method Putkisto, Alexander Technique, Feldenkrais, Hellerwork, Rolfing movement, Gabrielle Roth's 5 Rhythms, Traeger, Tai Chi, Chi Kung and Yoga are all forms of somatic education. They focus as much on the *way* you do what you do as on *what* you do, in contrast to the increasing mechanization of the fitness industry and the medical establishment. This is the client-centred, softer, more feminine approach of somatic education towards a more human future. In this model of the body, hard is no longer good, and harder better (although tone and strength are valued). It is the movement that muscles can generate and be put through, rather than their 'resting' size, and the quality and appropriateness of that movement that matter.

Somatics is a new field of specialization. In the new gym the objective is to widen and deepen the channel of life within us, enhancing our capacity to experience the fullness of life and bringing harmony and integration to ourselves in our totality: mind, body, spirit and emotions.

The new gym works from the inside out; the emphasis is on releasing tension, stretching short or tight tissues, relating the mind and body, self-understanding, self-care, self-acceptance, self-expression, posture and the quality rather than the quantity of what you've got.

Connection is the key word in the new gym; the opportunity to put yourself more fully into the driving seat. Somatic education is about connecting us to ourselves, mind to body and body-mind to spirit; yourself to others and to the planetary and universal context. You are part of the whole.

SOMATIC EDUCATION

The distinguishing characteristic of somatic education is the active participation of the client in the therapeutic process. The therapist/coach acts as guide and facilitator. The degree of this participation may differ greatly, due to many factors and variables. There is a lot of cross-over between different brands of somatic education as the whole field becomes more established and self-conscious. A recent International Somatics Congress included workshops and master classes from 79 accredited teachers from more than 60 distinct somatics schools, concerned with movement, breath, energy, emotion, expression, ergonomics, spirituality, sexuality, communication, integration, relaxation, learning, self-help, healing and personal growth issues, challenges and opportunities.

THE NEW EXERCISES

Below are some examples of the exercises and practices of the new gym, to give you a flavour of what is on offer. Swim, dance, bounce, walk. Above all, have *fun* exercising. Increase your freedom, mobility, expressiveness, fluidity, dexterity and joy.

Rebounding

Rebounding uses a small 40-inch-diameter mini-trampoline. NASA says it is the most efficient exercise device ever! Rebounding uses gravity directly, as you accelerate down onto the trampoline you effectively weigh more, and at the top of the bounce you are momentarily weightless. It is as if gravity is being turned up and down, exercising every cell in your body. The best small, round mats are made in Germany and the USA; watch out for very cheap Far Eastern copies, which are too tightly strung and fall apart with any kind of consistent use.

Little and often is better than marathon sessions; use the rebounder whilst the bath is running, or to move your energy between phone calls or writing/studying sessions. Use it to release tension in the shoulders and arms. Rebounding is unique in the way it combines all the elements of fitness training in one activity: strengthening, stretching and building stamina simultaneously, whilst also making you smile and supporting your structure – it is very hard to bounce any other way beside up and down, in the line of gravity!

Full-size trampolining is great too, but is much more hazardous, needing supervision and professional training. The majority of trampoline injuries occur when two people are bouncing together, so don't get carried away.

Pilates

Joseph H. Pilates, who was born in Germany (near Dusseldorf) in 1880 and who overcame rickets, asthma and rheumatic fever as a child, developed his unique exercise equipment whilst interned as an enemy alien in the UK during the First World War. He migrated to America in 1926 and set up in a studio in New York in the late 1920s to teach what he called contrology – a series of exercises he had invented using simple machines to increase postural control – effectively to strengthen the body without shortening it, which is the main problem with conventional exercise. His methods were taken up by many dancers, as Pilates exercise helps to increase suppleness (and possibly through convenience, as his studio was above a ballet school!)

Because of this accidental association with the dance community, the propagation of Pilates' ideas remained within the dance culture for many years, even though Joseph Pilates' work has, and was intended to have, a wide applicability. It remains one of the most conscious, safe and whole exercise systems there is, although it has suffered from the lack of an anatomically based language to relate it to other methods due to its dance-based history. More recently Pilates' teachers have been reaching out to a wider market with books and videos, and it is becoming much better known.

Example of Pilates exercise

Stomach stretch. Lie on your front on the floor, supporting your forehead with the backs of your hands (a similar position to that used for prone undulation, *see Chapter 11*). Take a breath, and as you are breathing out, pull up your belly and try to hold it off the floor for a count of 10 seconds. Breathe in; and repeat 10 times. You should isolate the abdominal muscles for this exercise, in other words try not to tense anything else. Try not to let your belly down even when breathing in.

(This exercise was taught to me by the late Avi Shoshana, who is fondly remembered for his gentle humour, charisma and warmth. I still hear his inimical Israeli accent: 'Sto-mack Strrretch!')

Method Putkisto

Finnish-born Marja Putkisto, originally working with her own challenge of a congenital hip dislocation, has evolved a method of deep stretching which goes far beyond the stretching conventionally thought of as sufficient in standard exercise systems. Taking the idea many degrees on from stretching merely as a preventative add-on, she makes it the centre of an intelligent and anatomically informed approach to overall health. She is the only exercise therapist I know who has methods to stretch the diaphragm directly and claims, rightly I believe, that a healthy diaphragm and correct breathing can actually improve posture, reversing the conventional wisdom of implied causality. It is the action of breathing that is central to making the Method Putkisto stretches work so deeply.

Marja works at the Finnish Opera house and the Sibelius University of Music, and has worked in London as well since 1990.

Example of Method Putkisto exercise

Bow stretch (supine version). The purpose of this exercise is to increase the elasticity of the diaphragm and to lengthen the waistline.

Lie down on your back, extend your legs and cross your right foot over your left. Extend your right arm over your head and place your left hand on the right side of your lower ribcage. Focus on your right side and on increasing the space between your pelvis and ribcage, creating a bow shape. Concentrate on breathing in on the right side of your lungs, underneath your left hand, allowing your ribs to expand sideways into your hand.

Once you have reached maximum inhalation, pause (after this pause you may be able to breathe in a little more!), then begin to breathe out until you reach your maximum exhalation point and pause again. Create the stretch by increasing the bow at the point of maximum inhalation and relaxing as you begin to breathe out, continuing the stretch by following the flow of your breathing. Continue for four cycles of breath, and then slowly switch to the other side. On completion return to the centre, bend the knees up one at a time and rest in that position before getting up. (The advanced version of the exercise is done standing, using the weight of the upper body to increase the stretch.)

Frisbee

Think of a frisbee as a flat ball. Anything you can do with a ball has a frisbee version, and more. It is vicarious flying. Buy the best frisbee you can find. A cheap frisbee is a travesty – it won't have the stability and forgiveness of a professional model and you'll spend all your time trying to figure out if it will fly at all. Even an expert has trouble making a child's disc fly.

Use your frisbee as a tool to discover the variety of movement that you have in your body, your wrists, legs, shoulders and back. Any way you can hold a frisbee it can be thrown. Develop both hands, catch with either in any position, whatever 'fits'. Live in the moment: if you try catching the disc between your legs when it arrives at eye level, you will be in trouble. Find an expert and play/copy. Imitation is the quickest way to learn. Learn about aerodynamics, how to work with the breeze, the angle of attack, precession. Discover airbrushing, tipping, mid-air attitude changes ('macking'), nail delay and other exotica.

Frisbee can be informal, unstructured, played alone, in pairs or groups.

Gabrielle Roth's 5-Rhythms Movement

The 5-rhythms movement practice has been developed by Gabrielle Roth over a lifetime of study in dance, theatre and the healing arts. It is an accessible form of movement that is both fun freestyle dance and, practised deeply, can become a source for artistic expression and profound healing. The 5-rhythms provides a structure for the exploration of your own unique dance. Each rhythm ('flowing', 'staccato', 'chaos', 'lyrical' and 'stillness') moves you physically in a different way, catalysing different aspects of your being.

This practice is useful to those who have barely danced as well as to experienced students of both movement and dance. No-one is too old to begin. No-one is too experienced to learn. No-one has the wrong shape of body to dance.

Roth has produced videos of her work to use at home, and CDs of her music to move to. You learn to follow the dancer inside you and your vocabulary of expression in movement and in life is extended. Each time you dance your movement becomes an invitation to release yourself from the fog of your daily affairs and to transform your life into living, breathing art.

> This work calls on you to discover and explore your own shamanic self. It doesn't require going out and buying feathers or even beating a drum.

> But it demands listening to the beat of our own heart, finding your own rhythm, singing your own blues, writing your own story, acting out your own fantasies and seeing your own visions. This is a contemporary, urban-primitive, Western-Zen, right-now trip. It's a journey into wholeness, an initiation into a shamanic perspective.
>
> Gabrielle Roth, *Maps to Ecstasy*.

Alexander Technique

F. Matthias Alexander (1869–1955) was a Tasmanian actor who discovered that his recurrent laryngitis was the result of tension and stiffness in his head and neck in relation to the rest of his body. He set up a three-way system of mirrors to observe himself and, through long and patient obsevation, he came to articulate principles that not only enabled him to solve his voice problems, but also had a profound influence on his general health and well-being.

As an instructional form of treatment, Alexander technique asks the pupil to work through three stages of movement retraining: awareness, inhibition and conscious control of aberrant movement patterns. Once the pupil recognizes dysfunctional patterns, he or she can be taught to consciously inhibit them and integrate more effective ways of moving.

Alexander work defines the head and neck as the primary locus of control for the body. According to this theory, compression of the head–neck–spine axis initiates and sustains other problems throughout the body.

Example of Alexander exercise

As you stand in front of the mirror at the basin in the morning, about to wash your face or brush your teeth – stop! Instead of bending forwards and shortening the front of you to reach the tap, look at yourself in the mirror and begin the movement with your face, so that you keep the upper body as a unit and flex the hip joints. Don't be afraid to stick your bottom out a little!

The Feldenkrais System

Moshe Feldenkrais (1904–1984) was born in Russia. He emigrated to Palestine at the age of 13. Educated in France in the sciences, he was one of the first Europeans to earn a black belt in Judo. After crippling knee injuries he taught himself to walk again and, in the process, developed an extraordinary system for accessing the power of the central nervous system to improve human functioning.

The Feldenkrais system involves movement education designed to reprogramme the brain, which has caused the body to develop compensatory movement abnormalities in response to traumas the body has gone through. The Feldenkrais system evolved out of the observation that when aberrant compensatory movement develops in response to injury, re-injury becomes more likely. (This observation is borne out by independent research which suggests that 25 per cent of sports injuries are re-injuries.)

Reprogramming the brain to improve bodymechanics is seen as an integrative process. The technique takes place in two stages. The first is an experiential 'awareness through movement' stage, where the subject actively follows a series of verbal commands designed to weaken old movement patterns and establish new ones. The second stage – 'functional integration' – consists of a series of educational hands-on passive movements.

Example of Feldenkrais exercise

This is only a 'sample' movement, to illustrate the principle behind the Feldenkrais system. It is not designed to be of therapeutic value.

Slowly turn your head to the left, whilst simultaneously looking towards the right. Then return to the centre, and turn your head to the right whilst looking to the left.

Trager

Milton Trager (1910–1997) was a gymnast and boxer, who discovered his power of touch when he gave his trainer a rub-down. He later became a physical therapist and an M.D., but never lost his belief in, and use of, the simple art of touching.

Trager movement uses gentle, passive motions, concentrating on traction, rotations, oscillations and rocking. The intention of these movements is to alter the client's neurophysiology, and to create deep relaxation. The passive movements are confined to the mid-range of the joint's normal range of motion, with the integration of gentle spinal traction. The active element of Trager – 'mentastics' – takes the form of neuromuscular re-education through exercises similar in principle to Feldenkrais work.

Aston Patterning

Judith Aston, a teacher of dance, physical education and theatre, trained with Dr. Rolf (*see below*) and developed Rolf movement work. Her own work, Aston patterning, views human structure and function as assymetrical, unique in shape and tension patterns, with movement occurring naturally in three-dimensional spirals. As a treatment programme, Aston patterning combines visual observation of postural alignment and function with palpation of soft tissue to analyse restrictions that limit movement options. The approach includes evaluation before and after treatment, the creation of a body-map indicating hypo- and hyper-tense areas, bodywork, movement education and an ergonomic consultation.

Rolfing

Rolfing structural integration was created by Dr. Ida P. Rolf (1896–1979), a biochemist/physiologist with an interest in Yoga and knowledge of Alexander, osteopathy and chiropractic, who developed a treatment strategy to alter mechanically postural alignment in gravity. Dr. Rolf believed that structure determined function and that the connective tissue system determines structural alignment, developing a method of altering the shape and balance of a body using deep pressure and chest movement.

Rolfing is a mechanical intervention that attempts to change the histology and biomechanics of connective tissue via appropriate pressure on the myofascia. It is a standardized therapeutic process that takes place in consecutive periods of deep manual intervention. This technique is performed independent of specific diagnosis and complaint; its intention to restore integrity of length and tone to the myofascial system.

Hellerwork

Hellerwork is a treatment method that evolved directly from Rolfing. Joseph Heller (1940–), a mathematician and aerospace engineer was trained by and worked closely with both Dr. Rolf and Judith Aston. Hellerwork involves an extensive evaluation of postural alignment and movement, as well as the balance of myofascial tone and length across all the major joints. Hellerwork also attempts to integrate the attitudinal and psychological aspects of posture and movement. Through both dialogue and movement education, Hellerwork's intention is to educate and assist the client in uncovering attitudes and unconscious beliefs that contribute to and limit postural integrity and efficient movement dynamics. (*See also Introduction: provenance of Hellerwork.*)

Moving slowly is a way of engaging the intrinsic, deep muscles. Tai Chi's effectiveness can be partly explained by the fact that the slow movements engage the core structure and intrinsic musculature in a way which does not happen in other, more 'speedy' activities.

Slow movements can also help overcome repetitive strain injury (RSI), which is the consequence of trying to do subtle manipulative and dextrous work without consciousness of the deeper sources of effortless movement. Sharon Butler, a Hellerwork practitioner on the East Coast of America, has developed a set of gentle stretches which have proved highly effective for the treatment of RSI. Knowledge of her programme has spread quickly via the internet, because of the number of people who need help after long stints of keyboard work (not to mention mouse use surfing the net!). She is now taking her work into American corporations, teaching occupation-specific injury prevention programmes and saving the companies money on their workers' compensation costs, not to mention saving the workers the pain and stress of injury.

SUMMARY

- In somatic education a person learns to change physiologically through a process whereby change can become self-initiated and self-controlled.
- Somatic education doesn't just cover bodywork, but tackles all aspects of being human – an integrated mind, body and spirit workout.
- The *way* you do things is as important as *what* you do.
- Somatic Education Therapy is a fast-growing field with many opportunities and applications.
- The new exercises include a holistic viewpoint.

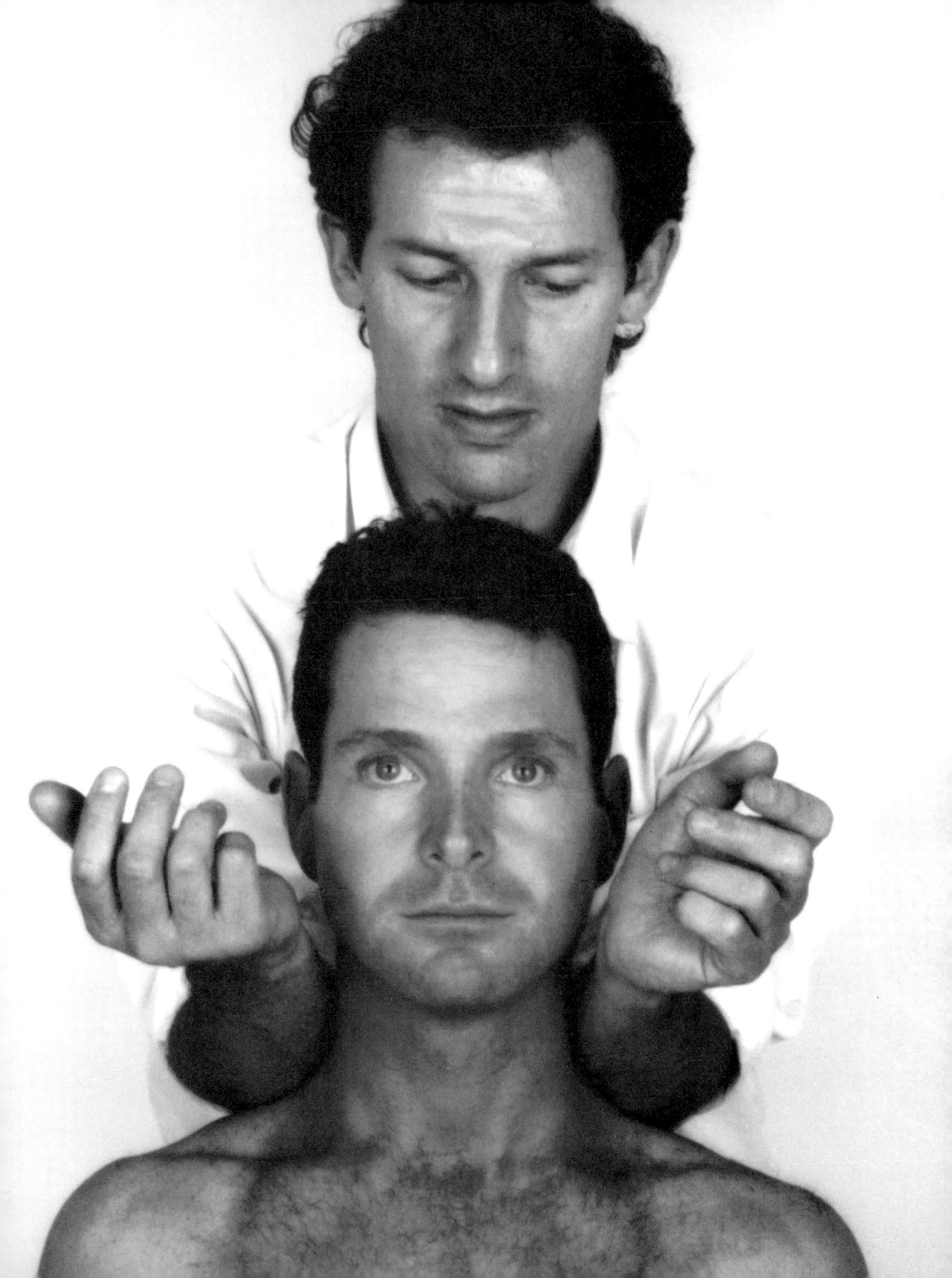

Chapter 14

The Ultimate Summary

ROGER GOLTEN'S SEVEN NO-NONSENSE LAWS FOR A HEALTHY, HAPPY LIFE

1) Don't forget to breathe – inspire. Fear is excitement without breath (Fritz Perls).
2) Be unreasonable – all progress depends on it (George Bernard Shaw).
3) Never wear uncomfortable shoes (Roger Golten).
4) Eat and drink consciously. Eat to live. Drink plenty of pure water (at least four pints per day). Make sure you eat good nutritional food – read labels for information on mineral/vitamin/efa/eaa/anti-oxidant content. 'Eat Organic, Save the Planet'(Craig Sams).
5) Connect with yourself. Find ways to centre yourself; meditate, get into bodywork, love yourself, forgive yourself, be kind to yourself. Get enough sleep. Remember you are a magnificent contribution (Erhard).
6) Connect with the Earth/Nature/Universe. Ground yourself – walk in nature, go camping or fishing, pick up litter, garden, swim, walk barefoot on sand or grass.

7) Connect with others. Share your heartfelt feelings. Listen with compassion in relationship, community and group. Apologize (Lew Epstein). Respect all life and points of view. Hug someone today (Virginia Satir). Make no judgements, avoid comparison and delete the need to understand (Brugh Joy).

References and Bibliography

This list contains books mentioned in the text, and selected books written by authors mentioned in the text. Out of print books or secondary sources are not mentioned, unless no alternative is available.

Alexander, F.M. *The Alexander Technique: The essential writings of F. Matthias Alexander*. Lyle Stuart, 1989.

Alexander, F.M. *Constructive Conscious Control of the Individual*. Larson Publications, 1997, abridged edition.

Batmanghelidj, Fereydoon. *Your Body's Many Cries for Water: You are not sick, you are thirsty!* Global Health Solutions, 1995.

Bond, Mary. *Balancing your Body: A self-help approach to Rolfing Movement*. Inner Traditions, 1996.

Clarke, Arthur, C. *3001*. Del Rey, 1997.

Dali, Salvador. *The Secret Life of Salvador Dali*. Dover Publications, 1993.

Epstein, Lew. *Trusting you are Loved: Practices for partnership*. The Partnership Foundation, 1998.

Feldenkrais, Moshe. *Awareness Through Movement: Health exercises for personal growth*. HarperSanFrancisco, 1991.

Fuller, R. Buckminster. *Critical Path*. St. Martin's Press, 1981.

Heller, Joseph and Henkin, William. *Bodywise*. Wingbow Press, CA, 1986.

Hoyle, Fred. *Home is Where the Wind Blows: Chapters from a cosmologist's life*. Universal Science Books, 1994.

Illich, Ivan. *Limits to Medicine, Medical Nemesis: The expropriation of health*. Pelican, 1977.

Illich, Ivan. *Disabling Professions*. Marion Boyars, forthcoming.

Joy, W. Brugh. *Joy's Way: A map for the transformational journey*. J.P. Tarcher, 1979.

Juhan, Dean, *Jobs Body: a handbook for bodywork*. Station Hill Press, NY, 1987.

Kurtz, Ron. *The Body Reveals*. ASIN 0060666863, o/p.

Liskin, Jack. *Moving Medicine: The life and work of Milton Trager M.D.* Talman, 1996.

Macdonald, Ian. *Revolution in the Head: The Beatles records and the Sixties*. Pimlico/Random House, 1994.

Morgan, Elaine. *The Aquatic Ape Hypothesis: The most credible theory of human evolution*. Souvenir Press, 1997.

Odent, Michel. *Entering the World: the de-medicalisation of childbirth*. Marion Boyars, 1989.

Pert, Candace. *Molecules of Emotion: The science behind mind-body medicine*. Simon & Schuster, forthcoming.

Pilates, Joseph, H. *Your Health*. Presentation Dynamics, 1998, first published 1934.

Pilates, Joseph, H. *Return to Life through Contrology*. Presentation Dynamics, 1998, first published 1945.

Putkisto, Marja. *Method Putkisto: Deep stretch your way to a firmer, leaner body*. Headline, 1997.

Robbins, Tom. *Even Cowgirls Get the Blues.* Bantam, 1990.

Rolf, Ida P. *Rolfing: The integration of human structures.* Harper & Row, 1977.

Roth, Gabrielle. *Maps to Ecstasy: Teachings of an urban shaman.* Nataraj Publications, 1992.

Roth, Gabrielle. *Sweat your Prayers: Movement as spiritual practice.* Tarcher/Putnam, 1998.

Schultz, R. Louis and Feitis, Rosemary. *The Endless Web: Fascial anatomy and physical reality.* North Atlantic Books, CA, 1996.

Shaw, Steven. *The Art of Swimming: In a new direction with the Alexander Technique.* Ashgrove, 1997.

Sidenbladh, Erik. *Water Babies: Igor Tjarkovsky and his methods of delivering and training children in water.* A.& C. Black, 1983.

Sky, Michael. *Breathing: Expanding your power and energy.* Bear & Co., 1990.

Todd, Mabel Elsworth. *The Thinking Body: A study of the balancing forces of dynamic man.* Princeton Book Co., 1937.

Trager, Milton. *Movement as a Way to Agelessness: A guide to Trager Mentastics.* Talman Co., 1995.

Watson, Lyall with Derbyshire, Jerry. *The Water Planet: A celebration of the wonder of water.* ASIN 0517565048, o/p.

Useful Internet Addresses

Note: Email addresses are not quoted unless there is no website; telephone numbers are not given unless there is no website or email address! If you haven't got access to the internet at home, why not go to an internet cafe and have a go? It's so easy!

Abby Rockefeller – www.clivusmultrum.com
Alan Raddon (shoemaker) – alraddon@aol.com
Brugh Joy – www.brughjoy.com
Buckminster Fuller – www.bfi.org
Craig Sams – www.earthfoods.co.uk
F.M. Alexander – www.alexandertechnique.com

Fereydoon Batmanghelidj – www.watercure.com
Frisbee – www.frisbee.com
Gabrielle Roth – www.ravenrecording.com
Hellerwork – www.hellerwork.com
– www.hellerwork.co.uk
Ida P. Rolf – www.rolf.org
Inversion therapy – www.inversiontherapy.org
John C. Lilly – www.garage.co.jp/lilly
Joseph H. Pilates – www.pilates.net
Joseph Heller – www.hellerwork.com
Judith Aston – astonpat@aol.com
Lew Epstein – www.partnership.org
Marja Putkisto – mputkisto@compuserve.com
Mary Bond – meerybe@flash.net
Millennium Dome –
www.new-millennium-experience.co.uk
Milton Trager – www.trager.com
Moshe Feldenkrais – www.feldenkrais.com
Peter Opsvik – www.stokke-furniture.no
Rebirthing – www.thecosmicbreath.com
Rebounding – www.healthbounce.com
– www.needakmfg.com
Rebounding (UK) – +44 (0)171 284 1918
Roger Golten – www.golten.net
Sarah Black (trichologist) – roneyclinic@netscape.net